Lecture Notes in Energy

Volume 29

Lecture Notes in Energy (LNE) is a series that reports on new developments in the study of energy: from science and engineering to the analysis of energy policy. The series' scope includes but is not limited to, renewable and green energy, nuclear, fossil fuels and carbon capture, energy systems, energy storage and harvesting, batteries and fuel cells, power systems, energy efficiency, energy in buildings, energy policy, as well as energy-related topics in economics, management and transportation. Books published in LNE are original and timely and bridge between advanced textbooks and the forefront of research. Readers of LNE include postgraduate students and non-specialist researchers wishing to gain an accessible introduction to a field of research as well as professionals and researchers with a need for an up-to-date reference book on a well-defined topic. The series publishes single and multi-authored volumes as well as advanced textbooks.

More information about this series at http://www.springer.com/series/8874

Yosef Jabareen

The Risk City

Cities Countering Climate Change:
Emerging Planning Theories and Practices
around the World

 Springer

Yosef Jabareen
Faculty of Architecture and Town Planning
Technion: Israel Institute of Technology
Technion City, Haifa
Israel

ISSN 2195-1284   ISSN 2195-1292 (electronic)
Lecture Notes in Energy
ISBN 978-94-017-7913-5   ISBN 978-94-017-9768-9 (eBook)
DOI 10.1007/978-94-017-9768-9

Printed on acid-free paper

Springer Science+Business Media B.V. Dordrecht is part of Springer Science+Business Media
(www.springer.com)

To my little Warrd Jabareen

Acknowledgments

I would like to express warm gratitude to my colleagues at Harvard, MIT, Ben-Gurion University, and Tel-Aviv University who inspire and encourage me to write this book. I am deeply indebted to Jerold Kayden, who first brought me to Harvard University and who continues to provide me with sound advice. I would also like to thank the generous scholars at Harvard University and MIT who have had an important influence on my insights and my career, specifically: Diane Davis, Lawrence Vale, John de Monchaux, and Bishwapriya Sanyal. I am also grateful to Oren Yiftachel and Tali Hatuka for their unconditional support.

This book was written and published thanks to the valuable assistance of many. Of special importance has been the editorial team at Springer, particularly Mark de Jongh and Cindy Zitter. I am also grateful to Geremy Forman, for his valuable comments and editing, and to my research assistants Helly Hirsh, Natalie Mickey, and Semion Polinov.

Contents

List of Figures

List of Tables

Chapter 1
Introduction

Contemporary cities and their residents are currently facing phenomenal mounting levels of evolving risk and vulnerability stemming, *inter alia*, from social polarization, the growth of urban poverty levels, urban conflict and violence, terrorism, natural disasters, and, most recently, climate change. Cities have been contending with risks related to security and some aspects of environmental disasters since ancient times, and the intensive urbanization, growth, industrial development, and technological progress of the twentieth and early twenty-first century have compounded long-standing risks and uncertainties and created new ones. In recent years, tens of thousands of people have lost their lives as victims to these intensifying risks and threats, and the social and material infrastructures on which human life depends have been severely impacted in urban locations throughout the world. Scientists from various disciplines today agree that the destructive impact of the risk and uncertainties stemming from climate change (both those we can anticipate today and those which are as yet unknown) are likely to increase in the near future. Others argue that climate change, with its catastrophic impacts, is already happening.

In this book, I argue that risk is a constitutive concept of our cities and propose the conceptual framework of the "*risk city*" as a praxis with the potential to make a significant contribution to our understanding not only of risk and its social, spatial, structural, and physical impacts on our contemporary cities, but on the way in which cities cope with uncertainties and vulnerabilities. This book takes seriously the threats and the mounting risk facing our cities today and calls for a paradigm shift in the way we think about, interrogate, and approach urban planning. The contemporary conditions of risk stemming primarily from climate change and its resulting uncertainties, I posit, challenge the concepts, procedures, and scope of conventional approaches to urban planning, as well as our overall planning culture. The result is a pressing need to rethink and revise current planning theories and practices. Indeed, in light of the phenomenal challenges we now face, our cities simply cannot continue on with business as usual in this realm.

The aim of this book is to theorize the risk city, with a focus on the risks stemming from climate change, and to conceptualize its planning practices geared toward coping with risk and uncertainties (though not employed in this manner

© Springer Science+Business Media Dordrecht 2015
Y. Jabareen, *The Risk City*, Lecture Notes in Energy 29,
DOI 10.1007/978-94-017-9768-9_1

here, the risk city as a conceptual framework can also be used to understand risks other than those related to climate change). The following chapters conceptualize city resilience and propose a practical framework for achieving and assessing this important goal. To examine the ways in which cities cope with risk and uncertainties, we analyze master and strategic urban plans and practices from a sample of ten cities around the world: Amman, Barcelona, Beijing, Delhi, London, Moscow, New York, Paris, Rome, and Sao Paulo. Special emphasis is placed on our analysis of New York City's ambitious *PlaNYC 2030* and on Hurricane Sandy's subsequent impact on the city and the surrounding region, which provides us with a valuable opportunity to examine the effectiveness of the planning practices employed by the city to cope with its uncertainties.

1.1 Spatializing the Risk Society

Social scientists have focused their attention on the concept of risk to the society at large and have invested little thought in spatializing risk on the city level. Anthony Giddens and Ulrich Beck conceptualize both modernity and modern societies in terms of risk. Giddens (1999) views risk as inseparable from modernity and as the mobilizing dynamic of societies that are bent on change and determined to control their own destiny rather than leaving it to religion, tradition, or the vagaries of nature. Prior to the modern era, cultures possessed no concept of risk and "lived primarily in the past," invoking "ideas of fate, luck or the 'will of the gods' where we now tend to substitute risk."

Beck (1992) defines the Risk Society in terms of risks that emerged in the 1960s. "Modern society," he maintains, "has become a risk society in the sense that it is increasingly occupied with debating, preventing and managing risks that it itself has produced" (Beck 2005: 332). From his perspective, this was "an inescapable structural condition of advanced industrialization." For Beck, the concept of "risk" replaces the concept of "class" as "the principal inequality of modern society, because of how risk is reflexively defined by actors." The theory of the world risk society, however, maintains that modern societies are shaped by new kinds of risks and that their foundations are shaken by the worldwide anticipation of global catastrophes. Such perceptions of global risk are characterized by three features (Beck 2005: 334): (1) spatial, as reflected in the fact that many new risks do not recognize the borders of nation-states and other such entities; (2) temporal, as manifested in the long latency period that are characteristic of new risks (such as nuclear waste), making it impossible to effectively determine and limit their effects over time; and (3) social, as exhibited in the complexity of the problems and the length of the chains of effect, which means that it is no longer possible to determine causes and consequences with any degree of reliability (as in the case of financial crises).

The Risk Society becomes a grand narrative that must be dismantled and deconstructed before we can truly understand its consequences. By theorizing the *risk city*, I seek to adapt this general notion to smaller-scale contexts of modernity by shifting attention from the Risk Society as a whole to the very real risks present at the urban level. By doing so, I am attempting to spatialize contemporary emerging risk and uncertainties in the context of the city as a human habitat.

Contrary to the lack of spatiality and borders supported by Beck, I argue the necessity of spatializing the contemporary emerging risks stemming from climate change and environmental hazards (as well as global terrorism and the like) and of situating them in human spaces—mainly cities and urban communities. I also maintain what many city administrators have been learning in recent years: that in order to effectively cope with uncertainties and risks, cities need to become key actors in the process. Indeed, contemporary cities are beginning to emerge as major forces in critical areas such as human security, sustainability, and climate change. Refocusing our analysis on cities increases our chances of understanding specific risk phenomena and the actions required to deal with them. On this basis, in my quest for a praxis, "a synthesis of theory and practice in which each informs the other" (Hillier 2010: 4–5), that is adequate for contending with both risk and its oriented practices, the modern city offers the best setting in which to situate our inquiry. As a result, this shift has the potential to make a substantial contribution at both the practical and the theoretical levels.

I argue that, to a certain extent, cities have always been about coping with risk, as expressed in the following words penned by Aristotle more than two millennia ago: "Men come together in cities for security; they stay together for good life" (Blumenfeld 1969: 139). With the rapid development of technology and modernity, this aspect of cities has intensified greatly, as reflected in their increasing occupation with interrogating, estimating, preventing, managing, accepting, denying, and seeking to manipulate and cope with risks. Indeed, cities have been facing environmental, health, social, and security threats for centuries, and have always strived to reduce risks by means of various spatial, physical, social, and environmental measures.

Therefore, my aim in this book is to develop the theoretical framework of the risk city with the primary goal of filling a gap in the academic literature with a framework that not only theorizes urban risk and its uncertainty but also interrogates human risk-oriented planning practices and contributes to our understanding of the effect of these practices on urban social issues, particularly those related to social justice.

1.2 The *Risk City*: The Theoretical Framework

The theoretical framework of the risk city, which I propose in Chap. 2, is based on the three primary concepts of risk, trust, and practice. Overall, the framework of the risk city is like a plane of immanence—an "image of thought" with interconnected concepts. Although it is the coexistence and reciprocal relationships among these interlinking concepts that give meaning to the risk city, each concept plays its own unique role in the framework. In accordance with Deleuze and Guattari's (1991) approach to the term "concept," each concept of the risk city is "created as a function of problems" or related to a problem or problems (p. 18), and "has a *becoming*" and a relationship with other concepts situated in the same conceptual framework, or "plane."

Accordingly, I define the risk city as a construct of the interlinking concepts of risk, trust, and practice:

(a) **The risk city as a construct of risk**: Risk is the ontological foundation of the risk city. The risk city is first about knowledge regarding threats and future uncertainties that are related to but not limited to environmental and climate change. Indeed, risk regulates the present and future of cities by significantly contributing to the mobilization of urban society and its politics, as well as its practices of planning and development. One major feature of the risk city is its ontological foundations that are rooted in the restlessness of knowledge regarding the risks faced by cities. For the most part, knowledge regarding risk is questioned and challenged not only by the public but by the experts themselves, which means that the risk city exists in the shadow of unstable, challenged, and restless knowledge.

Risk is socially and culturally constructed, interpreted differently and manipulated by different people with different interests and different backgrounds. This means that knowledge regarding risk stands on inherently restless terrain. Consequently, each society has its own conceptions of the risk city based on its own understanding and interpretation of uncertainties, knowledge, political organization and values, political and market powers, and resources. Risk means different things to different people depending on their social, economic, and political capacities and their political allegiances and social conditions. Following Douglas and Wildavsky's (1982) pioneering work on risk perception, social scientists have argued that risk behaviors and perceptions can neither be understood nor analyzed outside the social and cultural contexts in which they evolve (Sommerfield et al. 2002). In this way, risk perception varies according to historical traditions and cultural beliefs, as well as political and administrative structures (Healy 2004; Jasanoff 1986, 1999; Rohrmann 2006). Since risk is "a virtual threat," as posited by November (2008), many individuals, urban communities, and policy makers may not regard some types of risk as a serious or urgent matter.

Risk is also about power and resource allocation, and risk conception is therefore a tool of political and social power in our cities. Because risk reduction and treatment entails resource allocation and consumption, politicians and other economic stakeholders typically appropriate the right to reframe risk. Who is it, then, who conceives risk, and who are their receivers and their target audience? Experts and scientists usually reframe risk settings as a science for our societies and urban communities, and powerful stakeholders typically hijack the right to reframe the acceptable level of risk. Without a doubt, decision makers and politicians prioritize risk based on political, economic, and social considerations. It would be naive to suggest that in their dealings with risk and the risk city, politicians and decision makers consider scientific facts alone. According to Beck (1992, 2005), "even the most restrained and moderate objectivist account of risk implications involves a hidden politics, ethics and morality." The risk city involves social conflict on local and national levels and is also responsive and reflexive to the international politics and tension along the climate change divide in world politics. After all, as Beck reminds us, "not all actors really benefit from the reflexivity of risk—only those with real scope to define their own risks."

(b) **The risk city as a construct of trust**: Ultimately, the risk city seeks to promote trust and a sense of safety among its inhabitants and visitors by producing social and political institutional frameworks and promoting practices aimed at reducing risk and the possibility of risk. Therefore, the risk city negotiates, manipulates, and mobilizes trust whenever it approaches and deals with risk. The emergence of risk demands and is closely followed by the negotiation of trust. Trust is fundamental to the risk city because of its dialectical relationships with risk. For Giddens (1990: 35), "risk and trust intertwine, trust normally serving to reduce or minimize the dangers to which particular types of activity are subject." Trust can be defined as positive expectations in the face of the uncertainty emerging from social relations and from the relations between the citizenry and the authorities (Guseva and Akos 2001).

Like risk, trust is socially and culturally constructed. In the risk city, the feelings and perceptions of trust held by different individuals and social groups differ in quality and intensity. I also posit that different cities are characterized by different conceptions of trust based on their social structure, diversity, and demographic and socioeconomic conditions. Social trust is based on judgments of "cultural values," as individuals tend to trust institutions that, in their judgment, operate according to values that match (or are similar to) their own. These values vary over time, according to social context, and among individuals and cultural groups (Cvetkovich and Winter 2003). The literature on trust has shown variation across countries, between native-born and immigrant groups on the neighborhood level, and among cities. The absence of trust in the risk city has undesired consequences, leading to community and social disorganization which, in turn, increases crime and delinquency rates and destroys the sense of community.

Trust is more than a feeling of safety. It is also about the confidence in a city, its public authorities, and its physical and abstract settings. It is about trusting the city per se, and it plays a critical role in the risk city due to its social function of mitigating uncertainty. In the risk city, trust can emerge in a variety of forms, levels, and scales—from face-to-face exchanges and ascriptions to institutions, physical infrastructures, and technical systems. I maintain that in the risk city, the involvement of residents in planning and producing their own spaces increases the levels of trust among them.

(c) **The risk city as a construct of practice**: A primary aspect of the risk city is its construction of sociopolitical and spatial practices and frameworks aimed at responding to these uncertainties and countering the worst of them. In this way, it is about "structural arrangements," "emergency planning," prevention, mitigation, and adaptation. Both trust and risk help shape social practices in the risk city. Giddens (1976) uses the term "double hermeneutic" to refer to the observation that "when scientific concepts become generally accepted as means of making sense of the society, they not only reflect but also construct social practices" (Häkli 2009: 14). In this way, risk and trust not only describe but also construct social and planning practices related to the risk city.

The risk city, therefore, must be understood as a future-oriented socio-spatial political construct that dynamically mobilizes its various frameworks in an effort to determine its own future rather than leaving it to the hand of fate. The risk city makes positive use of risk conditions to creatively reconstruct itself and to address issues related to the people, energy, and environmental, spatial, and economic development. In the words of Beck (2005: 3), risk is "the modern approach to foresee and control the future consequences of human action."

The following problems with practice-oriented risk stem from the very essence of risk itself, particularly its underlying uncertainty and complexity: (1) Risk is future-oriented in nature and not an immediate or pressing need, and is therefore not typically treated as urgent and is often ignored for long periods of time. (2) Risk sometimes has to do with problems that are difficult, if not impossible, to understand and resolve scientifically. (3) Because the measures required to address risk may be costly, public authorities either ignore them or deal with them in a minimal manner. (4) As addressing risk sometimes offers no immediate political gain, many political leaders choose simply to quietly ignore it. (5) Practices that are informed by uncertainties are difficult to design and plan: although when addressing future risks planning must address complexity at the urban level, neither our extant urban theories nor our practical experience provide us with the adequate tools for producing such practices.

The risk city internalizes risk, trust, and practices, while each concept is internally contradictory by virtue of the multiple processes and heterogeneous components by which it is constituted. As a result, these concepts constitute the risk city as a process that is both contradictory and unstable—a contradictory "thing" that can be understood as the processes and relationships among the concepts that constitute it and which it internalizes. The uncertainty of the risk city shapes it as contradictory, conflictual, and unbalanced. Thus, in order to understand the risk city we must conceive it as an assemblage of interlinking spatial, political, economic, social, and cultural processes replete with contradictions, conflicts, and sources of power, that together provide crucial insight into the complexity of urban life and settings. Moreover, the constitutive concepts of the risk city are not static but rather under continuous evolution and in a constant state of process. "Every concept has a history," explains Deleuze and Guattari (1991). Risk, too, has a history, as do trust and planning practices.

Risk and trust contribute their own power in forging our cities and their socio-spatial construction. The risk city has its own form and specific geographies of fear, its own spatiality, and its own structure. Through its structure and spatial typologies, its socio-spatial segregation, its enclaves, and its fortifications we can learn to interpret the risk city and how different people in the city conceive differently of risk and trust.

1.3   The Risk City as a "Lack" and an "Illusion"

The risk city can be considered to be in a state of "lack," to use the terminology proposed by French psychoanalyst and thinker Jacques Lacan. Because it does not engage in practices to address all types of risk, many aspects of the trust perceptions among city residents go unsatisfied. In this sense, the risk city seeks to provide what people feel they are lacking. I concur with Gunder and Hillier (2009), who introduce the concept of lack to spatial planning. Accordingly, the practices and plans of the risk city are believed to reduce doubt and uncertainty and to promise certainty in the future. However, these 'imaginary' "plans and their prescribed solutions lack" (Gunder and Hillier 2009: 29). The risk city, which lives upon unstable foundations, asks us to "continue to plan for certainty, even if we know—in our heart—that it is merely illusion and rationalization" (Gunder and Hillier 2009: 29).

The risk city does not tackle all types of risk. In addition to the *targeted risk* that it seeks to address and to mitigate there is also *accepted risk*, which the risk city accepts and with which it agrees to live without challenging (for various political, economic, and cultural reasons), and *ignored risk*, which the risk city either consciously or subconsciously disregards. This has a direct impact on regions of trust in the city, as some are constructed and reinforced through specific practices and plans (both real and 'imaginary') while others go untended as a function of the risk city's decision to accept some risks and ignore others. There is an emptiness to the risk city left by the perceived lack of security, certainty, sustainability, and trust by which it is necessarily characterized. It strives to fill this gap through social practices and "pragmatic social construction," through utopian vision, and through efforts to generate a desirable state (see Laclau 2003).

As *a praxis* that links theory with practice, the risk city acts to acquire knowledge regarding future uncertainties and to construct socio-political and spatial frameworks aimed at responding to and countering these uncertainties. It dynamically mobilizes its various resources in an effort to determine its own future, making use of the conditions of risk in a positive manner to creatively reconstruct itself and to address people, energy, and environmental, spatial, and economic development. Lack is one of its primary driving forces, as it seeks to win the trust of its residents, but can be only partially successful at doing so. After all, like its constituent components of risk and trust, the risk city is socially and culturally constructed and means different things to different people in different social and political contexts.

1.4   Emerging Planning Practices Countering the Risk of Climate Change

One of the fundamental assumptions underlying the risk city is that change in risk perception can be expected to lead to change in trust perception, both of which inform and induce the need for new planning practices to meet the uncertain challenges. In this way, the more recent incarnations of the risk city, with its knowledge regarding threats, uncertainties, and vulnerabilities, dramatically challenge conventional planning theories and practices.

In recent years, we have become increasingly aware of the major risks and uncertainties that climate change poses to our cities and communities (IPCC 2007, 2014). Climate change is likely to effect the social, economic, ecological, and physical systems and assets of every city. It is expected to result in higher temperatures and more intense rainstorms, droughts, and heat waves, all of which threaten to increase strains on materials and equipment, create higher peak electricity loads and voltage fluctuations, disrupt transport, and escalate the need for emergency management (Barnett 2001; Leichenko 2011; Peltonen 2006). Water supply and quality may also be affected, and energy transmission and distribution may be dramatically disrupted. Cities may also experience massive in-migration from affected rural areas pressured by drought or other climate extremes. Climate change will also affect urban security and threaten the well-being, safety, and survival of urban people (see Barnett and Adger 2005; Crawford et al. 2015; Rosenzweig et al. 2011). On a financial level, it will be costly to cities and states, as reflected in the costs of Hurricane Katrina, which was estimated to have caused over $100 billion (NOAA 2011) in damage. Finally, the effects of climate change will continue to deepen and broaden poverty among various urban low income groups.

In this way, climate change and its resulting uncertainties challenge the concepts, procedures, and scope of conventional approaches to planning, creating a need to rethink and revise current planning methods. Truly, when it comes to cities, "we are still at the stage of setting agendas and directions for research, and there are more questions than answers" (Priemus and Rietveld 2009: 425). Harriet (2010: 20) asserts that "we simply do not know what the impact of many of the initiatives that have been undertaken over the past two decades has been or what these achievements might amount to collectively." Countering climate change in cities is a complex and multidisciplinary phenomenon that demands a "paradigm shift" toward transdisciplinary thinking. Yet, most of the literature on the subject is fragmented and fractional in scope and typically overlooks the multidisciplinary nature of the subject. Without a doubt, focusing on only one or a limited number of aspects ultimately results in partial or inaccurate conclusions, misrepresentation of the multiple causes of the impacts of climate change, ineffective policy, and "unfortunate and sometimes disastrous unintended consequences" (Bettencourt and Geoffrey 2010).

Because risk entails uncertainty, it is a phenomenon with which planning has thus far failed to effectively cope. For decades, planning theories, practices, and education have been dominated by linearity. The same is true of the manner in which planning practices have contended with urban problems and threats in even the most advanced cities. Planning demands an appropriate approach to complexity, which is something that has yet to emerge and that must be addressed immediately.

This book seeks to understand and theorize the practices of the risk city as they relate to the risk and uncertainties with the empirical focus on climate change and cities. In Chap. 3, I propose *Planning for Countering Climate Change* (PCCC) as a framework for enabling our contemporary risk cities to counter climate change in the best manner possible. The development of this framework is based on the theorization of practices around the world as well as on an analysis of the inter-disciplinary literature on climate change in general and at the urban level in particular.

PCCC as a praxis, or a theoretical-practical framework of the risk city, is geared toward reducing the risk and uncertainties stemming from climate change. PCCC is an emerging approach for planning contemporary cities aiming at countering cli-mate change impacts, adapting cities to future uncertainties, and protecting resi-dents from environmental hazards and risk.

The PCCC framework is comprised of the following six concepts, each of which draws on different content and is rooted in different theoretical settings: mitigation, adaptation, equity, integrative urban governance, ecological-economics, and uto-pian vision. Although the concepts of mitigation and adaptation are dominant in the literature on environmental and ecological climate change, these criteria or concepts are theoretically and practically insufficient for providing a full account of the praxis of the risk city from the perspective of climate change. For this reason, the other four constitutive concepts are integrated to facilitate understanding of the theoretical and practical settings of the risk city. The concept of equity represents the elements of justice and ethicality that are related to climate change oriented practices, whereas integrative urban governance suggests how to manage the risk city through integration among institutions and the development of new organi-zational capacities to meet the challenge of the risk of climate change. Integrating ecological economics, or the "green economy," into the practices of the risk city helps facilitate and promote more effective environmental practices. Finally, the vision articulated by these practices—the vision of the risk city—reframes the problem and the lack of the current situation and calls for filling this lack and the existing gaps in the present or in the future.

The main features of PCCC reflect that this framework differs from conventional planning methods in approach, data analysis, visioning, and procedures. A major premise of PCCC is that planners must also think in terms of urban defensibility, or protection of the city, and therefore consider the state of critical infrastructure and the possibility of providing protection through new measures (such as natural infrastructure projects and coastal ecosystem restoration to create additional lines of storm defenses). In PCCC, future risk and uncertainties contribute to spatial plan-ning and, subsequently, to the location of new developments and growth patterns that avoid areas with high vulnerability. PCCC seeks to ensure alternative func-tional routes and infrastructures in the event of an extreme event.

Climate change played a central role in formulating the problem, in visioning and goal setting, and in the outcomes. At its core, PCCC is based not only on demographic, economic, and spatial analysis but also on the analysis of risk and uncertainties. Knowledge regarding climate change impact has become a critical resource for spatial planning. PCCC identifies human spaces, places, and assets that are vulnerable to extreme weather events, storm surges, sea-level rises, temperature changes, seismic events, and other such weather-related phenomena. In addition,

PCCC employs the Urban Vulnerability Matrix in order to more thoroughly understand the social-spatial distribution of risk and uncertainties. In this context, it operates to address the threats posed at the specific level of communities and social groups. Public involvement and participation is also of critical significance for PCCC, which, unlike conventional planning approaches, also incorporates adaptation measures.

Also unlike conventional planning approaches, PCCC incorporates energy as a major guiding concept for planning cities and communities, approaches land use based on the analysis of data related to risk and uncertainties, and integrates the development of scenario-planning capability as part of its procedures and outcomes. Overall, PCCC is easy to grasp and has the potential to facilitate greater awareness among scholars, professionals, decision makers, and the public as a whole regarding the current and future direction of cities in the arena of climate change issues. As a multifaceted conceptual framework, PCCC can help us determine what needs to be done to increase the resilience of our cities and, in this way, enable us to work more effectively toward making them more safe and secure.

1.5 Practices Around the World

In Chap. 4, I present a multifaceted conceptual framework for evaluating urban plans from the perspective of coping with climate change: the *Countering Climate Change Evaluation Method*. This evaluation framework is based on the theorization of practices of PCCC, presented in Chap. 3. Chapter 5 uses the PCCC assessment methods to examine recently issued inclusive, master, strategic, and climate change action plans of ten cities around the world (Paris, London, New York City, Amman, Rome, Sao Paulo, Delhi, Beijing, Moscow, and Barcelona), providing us with a valuable opportunity to compare the risk perceptions, the approaches to reducing risk and countering climate change, and the practical measures proposed by different cities around the world, from highly developed metropolises to less developed urban areas.

The analysis reveals that city plans have become exceedingly significant instruments of the risk city for coping with risk in general and risk stemming from climate change in particular. Plans are powerful instruments because they can bring mitigation, adaptation, and social, economic, and spatial measures and policies into integrated focus under a single plan. One of my main conclusions in this context has been that spatial planning is essential in the efforts of cities to cope with risk and threats. Moreover, city planning and development have an important role to play in contending with the future impact of climate change, the complexities and uncertainties of which pose new theoretical and practical challenges.

In the context of planning the risk city, it is clear that different perceptions of risk inform and induce different planning practices in different cities. Some cities have used their plans to articulate their view that one major risk with which they must cope is the risk stemming from climate change and environmental hazards. These cities have seized the opportunity presented by increasing knowledge and aware-ness regarding climate change to propose new inclusive plans for their cities. The resulting plans call for coping with climate change, and, at the same, for re-planning and restructuring the city and developing its social and economic spaces. These plans, which have been issued only recently and only by a handful of cities, promote a more inclusive planning approach that take climate change more seri-ously and further integrate spatial, social, and economic policies. Cities that take climate change seriously have applied a broad range of mitigation measures aimed at GHG emission reduction. Nonetheless, cities have been neither productive nor creative in the undertaking of adaptation. That is to say, notwithstanding a number of slight differences among the cities considered, they have all failed in their approaches to adaptation, forcing us to conclude that our cities are not doing all they can to fortify themselves and their residents against uncertainties, climate change, and natural and environmental hazards.

Other cities, which I assume are representative of the vast majority of cities around the world (in Russia, China, and other developing countries), appear to perceive risk differently: that is, as related not to climate change but to future growth opportunities. Their primary concern is with expansion, economic devel-opment, and international competition. "Growth" is the buzz word of most of the plans formulated by developed and developing cities and the inherent mission of those who issued them, even those who regard climate change risk as a major concern. As a result, the plans of London, Beijing, Amman, Delhi, Moscow, and many others cast expansion and growth as a major concept of development for their cities. Cities with growth concerns that are not climate change-oriented but related to housing, urbanization, transportation, physical infrastructures, and restructuring continue to apply traditional modern planning approaches, by which I mean planning regarding land use, zoning, urban spatial expansion, transportation system expansion, the development of networks of roads for private vehicles, and the establishment of new industrial areas—all without integrating concepts of sus-tainability or climate change concerns. The plans of Amman, Moscow, and Beijing seek to enhance growth and economic development without serious consideration of environmental concerns and without the use of sustainable transportation plan-ning, green building, mitigation codes for new buildings, the renovation of existing buildings aimed at reducing the use of energy, and renewable energy. In this way, the recent plans of Amman, Moscow, Delhi, and Beijing resemble plans from the 1920s and the 1950s. Indeed, the vast majority of cities have ignored the sustainable planning approaches and measures that humanity has been busy developing over the last two decades (at least).

Significantly, all the plans have failed to effectively integrate civil society, communities, and grassroots organizations into the planning process. The lack of a systematic procedure for public participation throughout cities' neighborhoods and among different social groupings and other stakeholders is a critical shortcoming, particularly during the current age of climate change uncertainty.

To meet the challenges posed by climate change in the current context of unprecedented uncertainty, planners are in need of a more coordinated, holistic, and multidisciplinary approach. Few cities, however, have thus far invested effort in achieving integrative urban governance. In some countries, another factor resulting in the failure of cities to do so has been the existence of a highly centralized national political system, such as that of China and Russia. Ultimately, our cities are neither properly nor effectively fulfilling the critical role they should be playing in coping with the risk and uncertainties facing their own residents. For this reason, they may end up functioning as a deathtrap for millions of residents when disasters occur.

1.6 The Deficiencies of Climate Change Master Planning

In 2007, New York City released *PlaNYC 2030*, an ambitious landmark plan aimed at charting the city's future for the coming decades and addressing the challenges of climate change (Rosenzweig and Solecki 2010b; Rosan 2012; Solecki 2012). Chapter 6 examines this strategic plan and considers the role played by climate change in shaping the planning process and its various components, beginning with problem formulation and culminating in its many outcomes. Analysis of the plan indicates that climate change played a central role in formulating the plan's problem, justification, and visioning and objective settings. *PlaNYC* is a physically oriented plan focused primarily on reconstructing infrastructures, promoting greater compactness and density, enhancing mixed land use, sustainable transportation, greening, and renewal and utilization of empty parcels and brownfields. It applies an integrated planning approach, making use of the advantages of new urbanism, TOD, sustainable development, mitigation, and the monitoring of institutional policies. The plan also recommends efficient ways of using the city's natural capital assets, pays special attention to strategies for providing New York with cleaner and more reliable power, and creates a number of mechanisms for promoting its climate change goals and creating a cleaner environment for economic investment.

At the same time, however, *PlaNYC* inadequately addresses major social planning issues, such as social and environmental justice, diversity, poverty, and spatial segregation, which are crucial to New York City, the most diverse city in the world. It also fails to address the issues facing vulnerable communities due to climate change. The lack of a systematic procedure for public participation throughout the city's neighborhoods and among different social groupings and other stakeholders is a critical shortcoming, particularly during the current age of climate change uncertainty.

My main argument regarding PCCC is that rather than adapting conventional planning approaches, planning should be oriented toward dealing with uncertainties. To this end, planning must develop adaptation strategies for facing future uncertainties. *PlaNYC* is an economic development and infrastructure-oriented plan with deficient and inadequate adaptation measures and therefore failed to protect New York and its communities from the impacts of Hurricane Sandy in October 2012. Because the planning process did not involve adequate public participation, *PlaNYC* failed to understand the city's map of urban-community vulnerability and to effectively address the critical needs of various communities in facing the storm. For this reason, *PlaNYC* should be redrafted with an emphasis on implementing the lessons of Hurricane Sandy and on rethinking adaptation and resilience building measures for the city in general and for its poor and vulnerable communities in particular. As planning has the power to protect cities and save lives, planners should assume a leadership role and become more involved in fighting climate change on the city level.

1.7 The Risk City Resilience Trajectory

The risk city is future-oriented, as are the planning practices of the risk city. Chapter 8 offers insight into the future resilience of the risk city, or what I refer to as the *Risk City Resilience Trajectory*. The main question in this chapter is how resilient our cities are and how we can accurately anticipate their future trajectories based on present planning practices. I posit that a city's resilience is composed of four interlinking dimensions: social, economic, environmental, and security. This book focuses on environmental resilience, that is, resilience related to environmental crises and the impacts and threats of climate change. The idea underlying the Risk City Resilience Trajectory is that our cities must learn from the past and the present in order to plan for the uncertainties of the future, since "resilience requires frequent testing and evaluation" (NYS 2100 Commission 2013: 7). Learning should be based primarily on our experience and constantly evolving knowledge on vulnerability and adaptation measures. The Risk City Resilience Trajectory enables planners to acknowledge both current and future vulnerabilities and risks and to plan a different future. Or, to use the words of Judith Robin in the aftermath of Superstorm Sandy, it enables cities " to build back better and smarter" (NYS 2100 Commission 2013: 7).

The *Resilient City Framework*, which addresses the critical question of what actions should be taken by cities and their urban communities to move toward a more resilient future, is a network, or theoretical plane, of four related concepts that allows for a comprehensive assessment of city resilience. Each of the four constitutive concepts plays a specific role in the framework and has a specific domain. The first concept, the "Urban Vulnerability Matrix Analysis," focuses on the governance culture, processes, arenas, and roles of the resilient city and plays a critical role in the resilient city due to its contribution to the spatial and socioeconomic mapping of future risks and vulnerabilities. "Urban governance," the second concept, which helps facilitate the holistic management of urban resilience, focuses on urban policies and assumes a significant need for a new approach to urban governance in order to cope with uncertainties and future environmental and climate change impact challenges. The concept of urban governance suggests that the integrative governance approach, deliberative and communicative decision making measures, and ecological economics can have a great impact on moving our cities toward improved urban resiliency. The concept of "prevention" refers to the various components that must be considered to help prevent environmental hazards and climate change impacts (mitigation measures, adaptation of clean energy, and urban restructuring methods). The fourth concept, "uncertainty oriented planning," refers to planners' obligation to adapt their methods to help cities cope with uncertainties in the future. According to the Resilient City Framework, a city's resilience is measured by the overall ability of its governing, physical, economic, and social systems to learn and to plan and prepare for uncertainties and to resist, absorb, accommodate, and recover from the effects of a hazard in a timely and efficient manner, including the preservation and restoration of its essential basic structures and functions.

1.8 Climate Change Is "Already Happening": The Deficient Resilient City

Hurricane Sandy, which thrashed the East Coast of the United States between October 28 and 30, 2012, affected the areas of 24 U.S. states, including New York City. In light of the catastrophic damages this "superstorm" left in its wake, some have argued that climate change in New York City is "already happening" (Gibbs and Holloway 2013). The impact of the storm left forty-three New Yorkers dead and tens of thousands injured, temporarily dislocated, or displaced altogether (Gibbs and Holloway 2013: 1). Across the country as a whole, the hurricane destroyed thousands of homes, caused the cancellation of 19,729 flights, left some 4.8 million people in 15 states and the District of Columbia without electricity, and killed over one-hundred people (Llanos 2012). The storm damage is projected to reach $50 billion making it one of the costliest natural disasters in U.S. history (the National Hurricane Center ranks Hurricane Sandy the second costliest U.S. hurricane since 1900). In parts of New York City, the storm caused sea levels to rise by 13 feet, offering a glimpse into possible conditions in cities around the world if the worst climate change scenarios are realized (it is interesting to note that climate change scientists posit that by 2200, sea levels in New York and elsewhere may rise by the same amount) (Chertoff 2012). Even if Sandy was not caused by climate change, it provides a concrete illustration the possible consequences of this process (Chertoff 2012). In any even, scientists warn that this type of storm may be even stronger in the future, with fiercer winds and heavier rains (Plumer 2012).

In light of the devastating effects of Hurricane Sandy, New York City is one of the cities that have applied efforts and resources to prepare itself for environmental crises and the impacts of climate change. Chapter 9 assesses the city's resilience in coping with the storm. In this chapter, I propose a qualitative assessment method (which is based on the previous chapter and neither "rigorous" nor "positivist" in approach) to determine the level of loss and the recovery time that is acceptable when qualifying a city as "resilient." The method is straightforward and easily comprehensible for the general public, decision makers, politicians, and practitioners alike.

The main conclusions of this assessment suggest that although the City of New York has a plan for countering climate change impacts that it has begun implementing, the city still appears to be unable to cope with future serious climate impacts. As revealed by the damage caused by Hurricane Sandy, the plan's major critical deficiency appears to lie in its shortage of adaptation measures for coping with environmental hazards. Like many cities around the world, including the most pioneering among them, New York City still does not employ comprehensive and spatial planning in its fight against climate change (Kern and Alber 2008). As we see in Chap. 5, most cities appear to be using mitigation policies alone to address human sources of climate change and have failed to apply adaptation policies.

Unfortunately, Sandy reveals the lack of resilience of our current institutional and spatial settings and the undeniable fact that our cities have become risky places for their residents during hazardous events. Also relevant to the damage suffered by New York is the planning process's distinct lack of adequate public participation, which contributed to the city's low level of resilience, particularly in its most vulnerable neighborhoods and areas. According to Uken (2012), the storm highlighted the fragility of the aging American infrastructure, with an electricity network that is ranked below those of considerably poorer nations. The lack of resilience demonstrated by most American cities in confronting Hurricane Sandy raises the question of whether these cities are capable of facing the future challenges of climate change without implementing dramatic adaptation measures and related policies. Researchers suggest that due to global warming, the number of future hurricanes will "either decrease or remain essentially unchanged" overall, but that those that do form will likely be stronger, with fiercer winds and heavier rains (Plumer 2012). If New York remains ill-prepared to face these extreme hazards, the city's residents will be vulnerable to tremendous injury.

The disaster caused by Hurricane Sandy made New York City and New York State painfully aware of the need for adaptation policies and strategies (NYC 2013; NYS 2100 Commission 2013). New York City (NYC 2013) has proposed a roadmap of strategic steps to be taken on the city level to improve its ability to protect life and property in the face of the increasing risk of severe weather, to increase the city's overall preparedness, and to generate the building blocks for a thorough and organized response to extended emergency events that may impact thousands of New Yorkers (2013: 5). New York State's NYS 2100 Commission (2013) has also suggested certain adaptation strategies in response to Hurricane Sandy.

The city's critical task at the present is to prepare it for the uncertainties of the future. In this context, it is important to learn from the crisis of Hurricane Sandy in order to improve its resilience for all. In the words of the NYS 2100 Commission (2013: 7): "We live in a world of increasing volatility, where natural disasters that were once anticipated to occur every century now strike with alarming regularity… We cannot just restore what was there before—we have to build back better and smarter." The NYS 2100 Commission has acknowledged and expressed regret that both the city and the state had previously accepted high levels of risk, explaining that "in recent storms, including Irene and Sandy, we have successfully embraced the notion of 'failing safely,' accepting the inevitability of widespread disruptions and tucking in to protect our assets to the extent possible" (NYC 2100 Commission 2013: 7).

1.9 The Risk City and the Challenge to the Neoliberal Agenda

The risk city as a praxis has the capacity to challenge the neoliberal agenda of cities. This is mainly due to the nature of the risk city, which demands public practices and encourages planning aimed at reducing uncertainties and vulnerabilities. The risk city needs public intervention to enhance its resilience and to promote the trust and safety of its residents. As illustrated by the effects of Hurricane Sandy, the demand for such practices is sometimes a matter of life and death.

Urban neoliberalism has recently become a deeply entrenched element of public sector administration, providing greater impetus than democratic political steering for administrative efficiency, entrepreneurialism, and economic freedom (Sager 2012). The nexus between mobile investment capital and public entrepreneurialism engender neoliberal policies (Sager 2011) that assume that social problems have a market solution (Peck and Tickell 2002). In this way, it promotes the supremacy of the economy and market, fewer restrictions on businesses, and the withdrawal of the state from social and economic safety nets, as reflected in processes of deregulation, privatization, and the devolution of central government. The predominant results of these policies have been social polarization and uneven economic development (Harvey 2005). Neoliberalism is about a restructuring of the relationship between private capital owners and the state, which demand promoting a growth-first approach to urban development (Sager 2013: 130). In addition, the neoliberal agenda suggests minimizing public planning intervention or, in other words, asks the public to play a limited role in the risk city. The market, proponents of this approach believe, is capable of addressing risks and uncertainties, including the urban impacts of climate change.

Our analysis of the master plans of cities around the world reveals that the neoliberal agenda of growth, economic development, and expansion is currently extremely dominant in most countries, including the U.S., China, Jordan, Russia and India. This singular emphasis on growth and expansion, however, marginalizes the critical needs of the risk city. The planning approach reflected in *PlaNYC* is based on just such a neoliberal agenda, and therefore neglects major urban social and safety issues and focuses primarily on mobilizing economic development. In this way, what appears be a landmark plan for New York City ("the real estate capital of the world") actually uses the concept of "sustainability" primarily as a public relations effort to package and brand what is actually a strategic real estate growth plan (Angotti 2008a, b, c, d: 6). According to Marcuse (2008: 1), *PlaNYC* is about what Susan Fainstein has referred to as "the new urban renewal," "with its displacement of people and the primacy of narrow economic concerns."

The risk city cries out for appropriate and more sophisticated and advanced practices and approaches to planning that regard the trust and safety of urban communities as top priorities, even at the cost of less hysteric expansion and growth. Whereas neoliberalism seeks to create a "good business climate," the risk city should be seeking to render urban environments that are safer, less risky, and more just.

References

Angotti, T. (2008a). The past and future of sustainability June 9. In *Gotham gazette: The place for New York policy and politics*. http://www.gothamgazette.com.

Angotti, T. (2008b). Is New York's sustainability plan sustainable? Hunter College CCPD Sustainability Watch Working Paper. http://maxweber.hunter.cuny.edu/urban/resources/ccpd/Working1.pdf.

Angotti, T. (2008c). Is New York's sustainability plan sustainable? Paper presented to the Joint Conference of the Association of Collegiate Schools of Planning and Association of European Schools of Planning (ACSP/AESOP), Chicago.

Angotti, T. (2008d). *New York for sale: Community Planning Confronts Global Real Estate*. Cambridge, MA: The MIT Press.

Barnett, J. (2001). Adapting to climate change in pacific Island Countries: The problem of uncertainty. *World Development, 29*(6), 977–993.

Barnett, J., & Adger, N. (2005). *Security and climate change: Towards an improved understanding*. Paper presented at the Human Security and Climate Change Workshop, Oslo, June 21–23, 2005. http://www.gechs.org/downloads/holmen/Barnett_Adger.pdf.

Beck, U. (1992). *Risk society: Towards a new modernity*. London: Sage.

Beck, U. (2005). *Power in the global age*. Cambridge: Polity Press.

Bettencourt, L., & West, G. (2010). A unified theory of urban living. *Nature, 467*(7318), 912–913.

Blumenfeld, H. (1969). Criteria for judging the quality of the urban environment. In H. J. Schmandt & W. Bloomberg (Eds.), *The quality of urban life 3* (pp. 137–163). California: Sage Publications.

Chertoff, E. (2012). The Sandy storm surge: Is this what climate change will look like? The Atlantic, October 30. Retrieved December 2, 2014. http://www.theatlantic.com/technology/archive/2012/10/the-sandy-storm-surge-is-this-what-climate-change-will-look-like/264292/.

Crawford, A., Dazé, A., Hammill, A., Parry, J., & Zamudio, N. (2015). *Promoting Climate-Resilient Peacebuilding in Fragile States*. Geneva: International Institute for Sustainable Development (IISD).

Cvetkovich, G., & Winter, P. L. (2003). Trust and social representations of the management of threatened and endangered species. *Environment and Behavior, 35*, 286–307.

Deleuze, G., & Guattari, F. (1991). *What is philosophy?* New York: Columbia University Press.

Douglas, M. A., & Wildavsky, A. (1982). *Risk and culture: An essay on the selection of technological and environmental dangers*. Berkeley, CA: University of California Press.

Gibbs, L., & Holloway, C. (2013). *Hurricane Sandy after Action*. Report and Recommendations to Mayer Michael R. Bloomberg. New York City.

Giddens. (1976). *The rules of sociological method*. London: Hutchinson.

Giddens, A. (1990). *The consequences of modernity*. Stanford, California: Stanford University Press.

Giddens, A. (1999). *Runaway world: Risk*. Hong Kong: Reith Lectures.

Gunder, M., & Hillier, J. (2009). *Planning in ten words or less: A lacanian entanglement with spatial planning*. Ashgate: Farnham.

Guseva, A., & Rona-Tas, A. (2001). Uncertainty, risk, and trust: Russian and American credit card markets compared. *American Sociological Review, 66*(5), 623–646.

Häkli, J. (2009). Geographies of rust. In J. Häkli & C. Minca (Eds.), *Social capital and urban networks of trust* (pp. 13–35). Farnham: Ashgate.

Harriet, B. (2010). Cities and the governing of climate change. *Annual Review of Environment and Resources, 35*, 2.1–2.25.

Harvey, D. (2005). *A brief history of neoliberalism*. Oxford: Oxford University Press.

Healy, S. (2004). A "Post-Foundational" interpretation of risk: Risk as "perfermonace". *Journal of Risk Research, 7*(3), 227–296.

Hillier, J. (2010). Strategic navigation in an ocean of theoretical and practice complexity. In J. Hillier & P. Healey (Eds.), *The Ashgate research companion to planning theory: Conceptual challenges for spatial planning* (pp. 447–480).

IPCC—Intergovernmental Panel on Climate Change. (2007). *Climate change 2007: Fourth assessment report of the intergovernmental panel on climate change.* Cambridge, MA: Cambridge University Press.

IPCC—Intergovernmental Panel on Climate Change. (2014). *Climate change 2014: Impacts, adaptation, and vulnerability.* http://ipccwg2.gov/AR5/images/uploads/IPCC_WG2AR5_SPM_Approved.pdf.

Jasanoff, S. (1986). *Risk management and political culture: A comparative study of science in the policy context.* New York: Russell Sage Foundation.

Jasanoff, S. (1999). The songlines of risk. *Environmental Values, 8*(2), 135–152.

Kern, K., & Alber, G. (2008). Governing climate change in cities: Modes of urban climate governance in multi-level systems (Chap. 8). In *Competitive Cities and Climate Change, OECD Conference Proceedings, Milan, Italy* (pp. 171–196). Paris: OECD. October 9–10, 2008. http://www.oecd.org/dataoecd/54/63/42545036.pdf.

Laclau, E. (2003). Why do empty signifiers matter to politicians? In S. Zizek (Ed.), Jacques Lacan (Vol. III, pp. 305–313). London: Routledge.

Leichenko, R. (2011). Climate change and urban resilience. *Current Opinion in Environmental Sustainability, 3*(3), 164–168.

Llanos, M. (2012). Sandy's mammoth wake: 46 dead, millions without power, transit. *NBC News.* Available at http://www.nbcnews.com/id/49605748/ns/weather/#.VILgV4vTb9s.

Marcuse, P. (2008). PlaNYC Is Not a Plan and It Is Not for NYC. Accessed March 3, 2013. http://www.hunter.cuny.edu/ccpd/sustainability-watch.

New York City. (2013). NYC Hurricane Sandy after Action Report—May 2013. *Report and Recommendations to Mayor Michael R. Bloomberg.* NYC.

NOAA—National Oceanic and Atmospheric Administration, National Weather Service, U.S. Department of Commerce. (2011). *Extreme weather 2011: A year for the record books.* Available at http://www.noaa.gov/extreme2011.

Peck, J., & Tickell, A. (2002). Neoliberalizing space. *Antipode, 34,* 380–404.

Peltonen, T. (2006). Critical theoretical perspectives on international human resource management. In I. Björkman & G. Stuehl (Eds.), *Handbook of international human resource management research* (pp. 523–535). Cheltenham: Edward Elgar Publishing.

Plumer, B. (2012). Is Sandy the second-most destructive U.S. hurricane ever? Or not even top 10? *The Washington post.* on November 5, 2012.

Priemus, H., & Rietveld, P. (2009). Climate change, flood risk and spatial planning. *Built Environment, 35*(4), 425–431.

Rohrmann, B. (2006). *Cross-cultural comparison of risk perceptions: Research, results, relevance.* Presented at the ACERA/SRA Conference. http://www.acera.unimelb.edu.au.

Rosan, C. D. (2012). Can PlaNYC make New York City "greener and greater" for everyone?: Sustainability planning and the promise of environmental justice. *Local Environment, 17*(9), 959–976.

Rodin, J., & Rohaytn, F. G. (2013). *NYS 2100 commission: Recommendations to improve the strength and resilience of the empire state's infrastructure.*

Rosenzweig, C., & Solecki, W. (2010). New York City adaptation in context (Chap. 1). *Annals of the New York Academy of Sciences* (Issue: New York City Panel on Climate Change 2010 Report).

Rosenzweig, C., Solecki, W. D., Hammer, S. A., & Mehrotra S. (2011). *Climate change and cities: First assessment report of the urban climate change research network.* Cambridge University Press.

Sager, T. (2011). Neo-liberal urban planning policies: A literature survey 1990–2010, *Progress in Planning 76*(4), 147–199.

Sager, T. (2012). *Reviving critical planning theory: Dealing with pressure, neo-liberalism, and responsibility in communicative planning.* London and NYC: Routledge.

Sager, T. (2013). *Reviving critical planning theory: Dealing with pressure, neo-liberalism, and responsibility in communicative planning.* London and NYC: Routledge.

Solecki, W. (2012). Urban environmental challenges and climate change action in New York City. *Environment and Urbanization, 24,* 557–573.

Sommerfield, J., Kouyate, M. S., & Sauerborn, R. (2002). Perceptions of risk, vulnerability, and disease prevention in rural Burkina Faso: Implications for community-based health care and insurance. *Human Organization, 2,* 139–146.

Uken, M. (2012). Sandy zeigt, wie marode Amerikas Infrastruktur ist [Sandy shows how ailing America's infrastructure is] (in German). Zeit Online (Hamburg, Germany). Retrieved November 02, 2012.

Chapter 2
Theorizing the Risk City

2.1 From the World Risk Society to the Risk City

When interrogating the concept of risk, social scientists have focused their attention on society at large and have invested little thought in space, or spatial spaces. Anthony Giddens and Ulrich Beck conceptualize both modernity and modern societies as a function of risk. In some senses society, or the "risk society," becomes a grand narrative that must be dismantled and deconstructed before we can truly understand its consequences. By theorizing the *risk city*, I seek to adapt this general notion to smaller-scale contexts of modernity by shifting attention from the risk society as a whole to the very real risks present at the urban level. By doing so, I am attempting to *spatialize* contemporary emerging risk and uncertainties in the context of the city as a human habitat.

Giddens (1999) views risk as inseparable from modernity and as the mobilizing dynamic of societies that are bent on change and determined to control their own destiny rather than leaving it to religion, tradition, or the vagaries of nature. Prior to the modern era, cultures possessed no concept of risk and "lived primarily in the past," invoking "ideas of fate, luck or the 'will of the gods' where we now tend to substitute risk." Modernization did not do away with the traditional view altogether, and concepts such as fate, god's will, providence, and other mystic notions still play an influential role among some, albeit as superstitions in which many only partially believe and often adhere to in a somewhat embarrassed manner. Giddens argues that in modern, future-oriented societies interested in change, risk has replaced notions of this kind.

Beck (1992) defines the "risk society" in terms of risks that emerged in the 1960s. "Modern society," he maintains, "has become a risk society in the sense that it is increasingly occupied with debating, preventing and managing risks that it itself has produced" (Beck 2005: 332). From his perspective, this was "an inescapable structural condition of advanced industrialization." Central to his theory of the risk society are relations of "risk definition" based on the "power game." These

© Springer Science+Business Media Dordrecht 2015
Y. Jabareen, *The Risk City*, Lecture Notes in Energy 29,
DOI 10.1007/978-94-017-9768-9_2

relations of definition can be conceived as analogous to Marx's relations of production, with the inequalities of definition enabling powerful actors to maximize risks for "others" and minimize risks for "themselves." For Beck, the concept of "risk" replaces the concept of "class" as "the principal inequality of modern society, because of how risk is reflexively defined by actors." The theory of the world risk society, however, maintains that modern societies are shaped by new kinds of risks and that their foundations are shaken by the worldwide anticipation of global catastrophes. Such perceptions of global risk are characterized by three features (Beck 2005: 334): (1) spatial, as reflected in the fact that many new risks (such as climate change) do not recognize the borders of nation-states and other such entities; (2) temporal, as manifested in the long latency period that are characteristic of new risks (such as nuclear waste), making it impossible to effectively determine and limit their effects over time; and (3) social, as exhibited in the complexity of the problems and the length of the chains of effect, which means that it is no longer possible to determine causes and consequences with any degree of reliability (as in the case of financial crises).

Contrary to the lack of spatiality and borders supported by Beck, I argue the necessity of spatializing the contemporary emerging risks stemming from climate change and environmental hazards (as well as global terrorism and the like) and of situating them in human spaces such as cities, towns, and villages. I also maintain what many city administrators have been learning in recent years: that in order to effectively cope with uncertainties and risks, cities need to become key actors in the process. Indeed, contemporary cities are beginning to emerging as major forces in critical areas such as human security, sustainability, and climate change. Refocusing our analysis on cities increases our chances of understanding specific risk phenomena and the actions required to deal with them. On this basis, in my quest for *a praxis* that is adequate for contending with both risk and its oriented practices, the modern city offers the best setting in which to situate our inquiry. As a result, this shift has the potential to make a substantial contribution at both the practical and the theoretical levels.

I argue that, to a certain extent, cities have always been about coping with risk, as reflected in the following words penned by Aristotle more than two millennia ago: "Men come together in cities for security; they stay together for good life" (Blumenfeld 1969: 139). With the rapid development of technology and modernity, this aspect of cities has intensified greatly, as reflected in their increasing occupation with interrogating, estimating, preventing, managing, accepting, denying, and seeking to manipulate and cope with risks. Indeed, cities have been facing environmental, health, social, and security threats for centuries, and have always strived to reduce risks by means of various spatial, physical, social, and environmental measures.

Since the industrial revolution, cities have been living under a steadily mounting level of risk. This has included climatic and environmental challenges that are independent of climate change, such as the urban heat island effect, by which cities are generally warmer than their surrounding areas due to higher levels of heat absorption and air pollution, and existing climate extremes, such as hurricanes and typhoons (IPCC 2007; Rosenzweig et al. 2011; The World Bank 2011). In con-junction with these phenomena, the impact of climate change on specific cities will depend on the actual climate changes experienced (such as higher temperatures and increased rainfall) and may vary from place to place. Changes in climate, in turn, will present an array of short and long-term consequences for cities in realms such as human health, physical assets, economic activities, and social systems. Without a doubt, the intensifying crisis in global climate change provides the argument for a shift to cities with even greater currency. Climate change has also resulted in a resurrection of the concept of risk. Giddens (1999) holds that in light of the current climate change crisis, risk assumes a new and peculiar importance. "Risk was supposed to be a way of regulating the future, of normalizing it and bringing it under our dominion," he explains, although "things haven't turned out that way. Our very attempts to control the future tend to rebound upon us, forcing us to look for different ways of relating to uncertainty."

Because of their socio-spatial character and large populations, contemporary cities are more vulnerable to a variety of risks and also have the potential to become generators of new risks, such as failed infrastructure and services, environmental urban degradation, and increasing informal settlements, which make many urban inhabitants more vulnerable to natural hazards and risk (UNISDR 2010). According to Beck, risk may even increase rather than decrease with progress in technology and science.

Cities are also where the vast majority of humanity will live in the coming few decades, as the demography of today's cities continues to change at an unprece-dented rate. While only 29 % of the world's population lived in cities in 1950, today the figure has reached 50.5 %, and by 2050 is expected to reach 70 %. The urban population is currently increasing at a staggering rate of 1 million people per week (Nature 2010) and, in Europe alone, is expected to rise from 920 million to 1.1 billion between 2010 and 2030. By the middle of the 21st century, the total urban population of the developing world is expected to more than double, from 2.3 billion in 2025 to 5.3 billion in 2050 (Satterthwaite 2007). Some suggest that we are currently witnessing an urban renaissance or a resurgence of cities (Storper and Manville 2006).

The threats stemming from the "manufactured risk" of climate change have been intensified at the urban level, which makes focusing on cities even more critical. According to Giddens (1999: 2), there are two types of risk: external risk, which stems from aspects of nature such as bad harvests, floods, plagues, and famines; and "manufactured risk, which is created by the very impact of our developing knowledge upon the world". Manufactured risk refers to risk situations which we have very little historical experience in confronting, such as climate change and most environmental risks. Such risks are directly influenced by what Giddens refers to as "intensifying globalization." In recent years, we have begun worrying more about what we have done to nature. According to Giddens, "this marks the tran-sition from the predominance of external risk to that of manufactured risk."

As the concept of risk continues to play an increasing role in how societies on various scales understand themselves and their actions and plan for the future, and as cities begin to emerge as key players in contending with risks of different kinds, we propose the conceptualization of the risk city—which focuses on society, structure, and politics at the city level—as a tool to better understand the urban settings that will soon be home to most of the earth's population. My theoretical entity, then, is the city: the urban space.

2.2 The Risk City: The Framework

My aim in this book is to develop the theoretical framework of the risk city with the primary goal of filling a gap in the academic literature with a framework that not only theorizes urban risk and its uncertainty and interrogates human risk-oriented practices and planning policy, but also contributes to our understanding of the effect of these practices on urban social issues, particularly those related to social justice. My goal is a praxis that is "a synthesis of theory and practice in which each informs the other" (Hillier 2010: 4–5).

The theoretical framework of the risk city is based on the three primary concepts: risk, trust, and practice. Although it is the coexistence and reciprocal relationships among these interlinking concepts that give meaning to the risk city, each concept plays its own unique role in the framework. In accordance with Deleuze and Guattari's (1991) approach to the term "concept," each concept of the risk city is "created as a function of problems" or related to a problem or problems (p. 18), and "has a *becoming*" and a relationship with other concepts situated in the same conceptual framework, or "plane." As part of one conceptual framework, these concepts are linked to and support one another, articulate their respective problems, and come together to contend with the same problem. Overall, the framework of the risk city is like a plane, a plane of immanence, an "image of thought" with interconnected concepts. After all, concepts, like knowledge, have meaning only in relation to an image of thought to which it refers or a conceptual framework which it serves.

The risk city internalizes risk, trust, and practices, while each concept is internally contradictory by virtue of the multiple processes and heterogeneous components by which it is constituted (see Harvey 1996: 51). As a result, they constitute the risk city as a process that is both contradictory and unstable—a contradictory "thing" that can be understood as the processes and relationships among the concepts that constitute it and which it internalizes. By contradictory we mean "two or more internally related processes that are simultaneously supporting and undermining one another" (Ollman 1990: 49, in Harvey 1996: 52). The uncertainty of the risk city shapes it as contradictory, conflictual, and unbalanced. Thus, in order to understand the risk city we must conceive it as an assemblage of interlinking spatial, political, economic, social, and cultural processes replete with contradictions, conflicts, and sources of power, that together provide crucial insight into the complexity of urban life and settings.

Fig. 2.1 The risk city framework

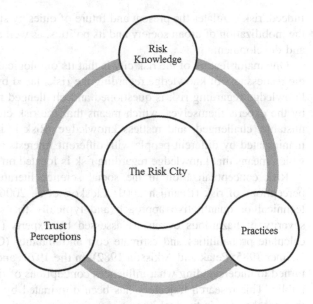

The constitutive concepts of the risk city are not static but rather under continuous evolution, in a constant state of process. "Every concept has a history," explain Deleuze and Guattari (1991). Risk, too, has a history, as do trust and planning practices. In fact, Giddens and Beck based their theory on the evolution of the concept of risk from traditional social contexts to the social contexts of modernity.

Based on the contemporary conditions of risk at the city level, I define the risk city as a construct of the interlinked concepts of risk, trust, and practice. The risk city is therefore a construct of *risk*, new evolving conditions, and knowledge regarding the uncertainties stemming from climate change and other perceived dangers. The risk city is also a construct of *practices* and practical frameworks employed by cities in response to the continually evolving and emerging knowledge about risk and uncertainties that stem from climate change and other threats. Finally, the risk city seeks to promote *trust* and a sense of safety among its inhabitants and visitors by producing social and political institutional frameworks and promoting practices aimed at reducing risk and the possibility of risk (Fig. 2.1).

2.2.1 The Risk City as a Construct of Risk

As we have seen, risk and its incessant uncertainties lie at the heart of the conceptualization of the risk city. In this way, risk is the ontological foundation of the framework. The risk city is first about knowledge regarding threats and future uncertainties that are related but not limited to environmental and climate change.

Indeed, risk regulates the present and future of cities by significantly contributing to the mobilization of urban society and its politics, as well as its practices of planning and development.

One major feature of the risk city is that its ontological foundations are rooted in the restlessness of knowledge regarding the risks faced by cities. For the most part, knowledge regarding risk is questioned and challenged not only by the public but by the experts themselves, which means that the risk city exists in the shadow of unstable, challenged, and restless knowledge. Risk is interpreted differently and manipulated by different people with different interests and different backgrounds, which means that knowledge regarding risk is located on inherently restless terrain.

Risk conceptualization in the social science literature is dominated by two perceptions of risk (Beamish 2001; Lidskog et al. 2006). The first is based on a technical or quantitative approach and typically focuses on the regularity and severity of hazardous events as assessed by experts (Jarvis 2007), as scholars calculate probabilities and estimate cost and liability (Crouch and Wilson 1982; Heimer 1985; Petak and Atkisson 1982). In the 1970s and 1980s, academic interest turned to understanding what influences conceptions of risk throughout the general public. This research trajectory has been dominated by a psychometric paradigm that emphasizes individual cognition using presumed "objective" measures (i.e., probabilities) of risk as the benchmark for comparison (Beamish 2001). From this perspective, risk is something that can be measured and observed. It is considered to exist when it can be assigned a probability of occurrence and is otherwise regarded as uncertainty (Gunder and Hillier 2009; November 2008). Renn and Rohrmann (2000: 14) define risk as the possibility of physical or social or financial harm/detriment/loss due to hazard within a particular time frame. 'Hazard' refers to a situation, event or substance that can be harmful for people, nature or human made facilities. 'People' at risk might be residents, employees in the workplace, consumers of potentially hazardous products, travelers/commuters or the society at large.

We expand this definition to include political harm as well, thus enabling risk analysis at the city level to accommodate political and religious conflicts and their real and possible outcomes.

The field of knowledge regarding climate change has been characterized by marked disagreement among scientists, as well as among political leaders and others from the private sector. While thousands of experts from different disciplines around the world highlight the risks, uncertainties, and effects of climate change, some scientists suggest that others are exaggerating the risks of climate change, or deny the phenomenon altogether. For example, in the wake of the Intergovernmental Panel on Climate Change's (IPCC) 2007 benchmark climate change report, the Inter Academy Council (IAC) undertook a review of the IPCC's findings that suggested that the 2007 report "contained exaggerated and false claims that Himalayan glaciers could melt by 2035." In this way, the IAC, which comprises many scientists, including some from the UK's Royal Society, issued "a damning report" into the research practices of the world's leading climate change body and called its credibility into question (Bowater 2010; IAC 2011).

The second perception of risk is social constructivist and qualitative in character. Following Douglas's and Wildavsky's (1982) pioneering work on risk perception, social scientists have argued that risk behaviors and perceptions can neither be understood nor analyzed outside the social and cultural contexts in which they evolve (Sommerfield et al. 2002; Jabareen 2006). Accordingly, some argue that understanding a person's interpretation of risk requires attention to the broader social, cultural, and historical contexts within which interpretation occurs (Beamish 2001; Erikson 1994). The social science literature also suggests that members of the same societal group are likely to adopt certain values and reject others, and this process of adoption and rejection is understood as determining the perceived acceptability of a risk (Snary 2004). In this way, risk perception varies according to historical traditions and cultural beliefs, as well as political and administrative structures (Healy 2004; Jasanoff 1986, 1999; Rohrmann 2006). The constructivist perspective also posits that risk "is not an objective condition, but a social construction of reality, which starts with the question of how people explain misfortune" (Hoogenboom and Ossewaarde 2005: 606).

Consequently, each society has its own conceptions of the risk city based on its own understanding and interpretation of uncertainties, knowledge, political organization and values, political and market powers, and resources. Risk means different things to different people depending on their social, economic, and political capacities and their political allegiances and social conditions. Overall, this helps us better understand different approaches to the risk city as well as differences in policies, planning, and development, and the agendas of sustainability and countering climate change. The underprivileged masses in developing cities, which are home to the vast majority of the earth's population, do not worry about global warming and species extinction, even though, they might suffer more than others from the impact of climate change. They conceive basic risk in different spheres using different terminology. For them, the most prevalent vocabulary for expressing risk conception pertains to food, access to clean water, employment, and urban hygiene.

One problematic aspect of the conceptual framework of the risk city is the fact that risk is "a virtual threat," as posited by November (2008). For this reason, many people, urban communities, cities, and policy makers may not regard the risk of climate change as a serious or urgent matter. From their perspective, many climate change oriented risks may simply not exist or may be conveniently ignored.

2.2.1.1 Power and the Conception of Risk

Risk is about power and resource allocation, and risk conception is a tool of political and social power in our cities. Because risk reduction and treatment entails resource allocation and consumption, politicians and other economic stakeholders typically confiscate the right to reframe risk. Who is it, then, who conceives risk, and who are their receivers and their target audience? Experts and scientists usually reframe risk settings as a science for our societies and urban communities, and powerful stakeholders typically hijack the right to reframe the acceptable level of risk. Without a doubt, decision makers and politicians prioritize risk based on political, economic, and social considerations. It would be naive to suggest that in their dealings with risk and the risk city, politicians and decision makers consider scientific facts alone. According to Beck (1992, 2005), "even the most restrained and moderate objectivist account of risk implications involves a hidden politics, ethics and morality." The risk city involves social conflict on local and national levels and is also responsive and reflexive to the international politics and tension along the climate change divide in world politics. After all, as Beck reminds us, "not all actors really benefit from the reflexivity of risk—only those with real scope to define their own risks."

Cities are willing to accept certain levels of risk and uncertainty. According to Giddens, the "acceptance of risk is also the condition of excitement," and "a positive embrace of risk is the very source of that energy which creates wealth in a modern economy." On the contrary, I believe that the acceptance of risks stemming from climate change and other such threats can be quite dangerous for the residents of cities. Indeed, recent evidence emerging from cities around the world suggests that it has simply become too dangerous to continue accepting the levels of risk that have previously been accepted in the arena of climate change.

2.2.2 The Risk City as a Construct of Trust

The risk city negotiates, manipulates, and mobilizes trust whenever it approaches and deals with risk. The emergence of risk demands and is closely followed by the negotiation of trust. Trust is fundamental to the risk city because of its dialectical relationships with risk. Theorists from various disciplines have emphasized the interrelationship between *trust* and *risk* (Beck 1996; Beck et al. 1994; Gambetta 1988; Giddens 1990, 1991; Kelley and Thibaut 1978; Josang and Presti 2004; Luhmann 1979; Molm et al. 2000). For Giddens (1990: 35), "risk and trust inter-twine, trust normally serving to reduce or minimize the dangers to which particular types of activity are subject." Molm et al. (2000: 1402) conceptualize trust as an emergent phenomenon that arises in response to uncertainty and risk. Trust can be defined as positive expectations in the face of the uncertainty emerging from social relations and from the relations between the citizenry and the authorities (Guseva and Akos 2001).

Trust reflects the social and political context of the risk city. It is more than a feeling of safety. It is also about the confidence in a city, its public authorities, and its physical and abstract settings. It is about trusting the city per se, and it plays a critical role in the risk city due to its social function of mitigating uncertainty. Trust is generally understood as a belief in the integrity of others (Ross et al. 2001; Guinnane 2005). Indeed, it is the fundamental bond of human society (Dunn 1984), and, according to John Locke, it is what "men live upon" (Locke 1976: 122). Barber defines trust as "socially learned and socially confirmed expectations that people have of each other, of the organizations and institutions in which they live, and of the natural and moral social orders that set the fundamental understandings for their lives" (Barber 1983: 165).

Residents of the risk city seek to or are supposed to trust the public authorities and institutions, as well as their various urban systems. In the risk city, trust can emerge in a variety of forms, levels, and scales—from face-to-face exchanges and ascriptions to institutions, physical infrastructures, and technical systems. We want to believe we can trust the subway system to work properly under all circumstances. Giddens (1990: 34) extends the definition to include "abstract principles" (such as technical knowledge) and institutions that relate to modernity and ultimately defines trust as "confidence in the reliability of a person or system, regarding a given set of outcomes or events, where that confidence expresses a faith in the probity or love of another, or in the correctness of abstract principles."

As a feeling that is central to human existence (Arrow 1972; Luhmann 1979) and a precondition for the existence of any society, trust fulfills important social functions in the risk city. Trust seems to make institutions, markets, and societies work better (Leigh 2006). It also promotes long-term social stability (Cook and Wall 1980), reduces the costs of exchange and transactions (Fukuyama 1995; Schmidt and Posner 1982), and enhances quality of life (Schindler and Thomas 1993). It is at once important for social exchange (Kollok 1994; Molm et al. 2000), an instrument of social control and protection (Barber 1983), and a vital component of social capital (Coleman 1988; Putnam 1995).

I maintain that in the risk city, the involvement of residents in planning and producing their own spaces increases the levels of trust among them. Studies have also found that trust also plays an important role in community development (Cebulla 2000; Dhesi 2000) and collaborative planning (Kumar and Paddison 2000).

Trust is related to the relationships among the public, the people, and their administrative and political institutions. Theoretically, public institutions seek to enhance trust in the face of uncertainty and emerging or anticipated risk. Practically, however, Gunder and Hillier (2009: 59) are correct in their assertion that "human societies increasingly reside in a life-world of fear and anxiety largely constituted by a loss of trust in our own ability and that of our national institutions to both ultimately know and deliver a better world."

Trust, like risk, is socially and culturally constructed. In the risk city, the feelings and perceptions of trust held by different individuals and social groups differ in quality and intensity. I also posit that different cities are characterized by different conceptions of trust based on its social structure, diversity, and demographic and socioeconomic conditions. Earle and Cvetkovich (1995) argue that social trust is based on judgments of "cultural values," as individuals tend to trust institutions that, in their judgment, operate according to values that match (or are similar to) their own. These values vary over time according to social context and among individuals and cultural groups. Cvetkovich and Winter (2003) have found a clear correlation between trust and assessments of shared salient values. The literature on trust has shown variation across countries (Knack and Keefer 1997; Uslaner 2002), between native-born and immigrant groups on the neighborhood level (Leigh 2006), and among cities within the United States (Alesina and La Ferrara 2005).

The absence of trust in the risk city has undesired consequences. Some studies suggest that the absence of trust leads to community and social disorganization, which, in turn, increases crime and delinquency rates (Sampson and Groves 1989; Shaw and McKay 1942).Others conclude that threatening conditions and high levels of disorder promote mistrust and destroy the sense of community (Greenberg and Schneider 1997; Ross et al. 2001; Skogan 1990; Taylor and Shumaker 1990). Convincingly, our study in Gaza Strip concludes that:

> Trust relationships are the corner-stone upon which communities are based anywhere in the world. Hence, in order to sustain communities, planners should support trust relationships among residents. This requires culture-sensitive planning (Jabareen and Carmon 2010: 446).

Moreover, elsewhere, I suggest that when it comes into being, risk frays the social fabric of the city, harms the fundamental roles of trust in the urban sphere and enhances the feelings of defenseless and vulnerability. Furthermore, it undermines social stability and quality of life, and harms co-operation among people, it destroys the sense of community and belonging and hurts trust in formal urban institutions and political legitimacy (Jabareen 2006b).

2.2.3 The Risk City as a Construct of Practice

Both trust and risk help shape social practices in the risk city. Giddens (1976) uses the term "double hermeneutic" to refer to the observation that "when scientific concepts become generally accepted as means of making sense of the society, they not only reflect but also construct social practices" (Häkli 2009: 14). In this way, risk and trust not only describe but also construct social and planning practices related to the risk city.

A primary aspect of the risk city is its construction of sociopolitical and spatial practices and frameworks aimed at responding to these uncertainties and countering the worst of them. In this way, it is about "structural arrangements," "emergency planning," prevention, mitigation, and adaptation. The risk city, therefore, must be understood as a future-oriented socio-spatial political construct that dynamically mobilizes its various frameworks in an effort to determine its own future rather than leaving it to the hand of fate. The risk city makes positive use of risk conditions to creatively reconstruct itself and to address issues related to the people, energy, and environmental, spatial, and economic development. In the words of Beck (2005: 3), risk is "the modern approach to foresee and control the future consequences of human action."

The field of practices of the risk city, of which countering climate change is but one, is a complex phenomenon that is non-linear, fundamentally non-deterministic, dynamic in structure, and uncertain in nature, and that is affected by a multiplicity of economic, social, spatial, and physical factors. Planning theories, however, provide us with no adequate solution for this complexity, and planners and practitioners lack the knowledge and experience necessary to deal with it properly, though many acknowledge the importance of this ability.

The problems with practice-oriented risk are related to the nature of risk itself, particularly its characteristic uncertainty and complexity:

(a) Because risk is future-oriented in nature and not an immediate or pressing need, it is typically not treated as urgent and is often ignored for long periods of time.
(b) Risk sometimes has to do with problems that are difficult, if not impossible, to understand and resolve scientifically (Bickerstaff et al. 2008: 1315).
(c) Because the measures required to address risks may be costly, public authorities either ignore them or deal with them in a minimal manner.
(d) Because in some cases addressing risk has no immediate political gain, many political leaders choose simply to quietly ignore it.
(e) Practices that are informed by uncertainties are difficult to design and plan. Although when addressing future risks, practices need to address complexity at the urban level, neither our extant urban theories nor our practical experience provides us with the adequate tools for producing such practices.

Because risk entails uncertainty (indeed, we have no idea when, where, and at what intensity it will actualize itself), it is a phenomenon with which planning has thus far failed to effectively cope. For decades, planning theories, practices, and education have been dominated by linearity. The same is true of the manner in which planning practices have contended with urban problems and threats in even the most advanced cities. Planning still lacks an appropriate approach to complexity, which is something that requires immediate elaboration.

Moreover, as climate change poses new risks and uncertainties that often lie outside our range of experience and that have the potential to affect the social, economic, ecological, and physical systems of any given city (see IPCC 2007: 719), countering climate change in cities is undoubtedly a complex and multidisciplinary undertaking that demands a "paradigm turn" toward interdisciplinary thinking. This, we must acknowledge, has yet to occur.

The concept of "complexity" offers a coherent perspective for organizing our knowledge in a variety of disciplines that has recently come to the forefront (Batty 2007). "Complexity sciences" is an interdisciplinary field concerned with the study of the general attributes of evolutionary natural and social systems (McGlade and Garnsy 2006) that evolved out of ideas associated with dynamic systems—ideas about chaos, nonlinearity, emergence, and surprise. Some argue that cities "are in the vanguard of these developments" (Batty 2007, 1; Marc and de Roo 2010: 93). Complexity driven research has placed a particular emphasis on "structural change driven by non-linear dynamics, as well as exploration of the propensity of complex systems to follow unstable and chaotic trajectories," and is increasingly viewed as an important step toward the construction of alternative evolutionary schemas (McGlade and Garnsy 2006, 1).

Social theorists borrowed the language of complexity from natural sciences (Urry 2005), and in the urban context, some scholars suggest that complexity thinking can add to our understanding of cities in general and cities of the twenty-first century in particular (Portugali 2010; Urry 2005). McGlade and Garnsy (2006) maintain that the emergence of alternative ways of representing relationships and the complexity of things has also opened up new possibilities in the social sciences, which have been dominated by the search for linear and predictive relationships that require heroic assumptions "that may distort rather than clarify." Complexity acknowledges "unpredictability and uncertainty, ambiguity and pluralism, and without being entirely relativist, it does throw doubt on the certainty of theory and science that has dominated our thinking about cities" (Batty 2007, 31).

Current approaches to complexity, however, are problematic in the sense that they are primarily quantitative and based on complicated computerized models, while urban phenomena are mainly qualitative in nature. These approaches "do not lend themselves to quantitative-statistical analysis and are thus of little interest to mainstream city complexity theory" (Portugali 2010). Although some qualitative urban phenomena can and have been modeled and simulated by means of complexity, urban phenomena, like the other burning questions facing the cities of the twenty-first century, are all "qualitative," with no hard data. For this reason, they are not addressed by the mainstream discourse of the complexity theories of cities (Portugali 2010). The potential contribution of complexity theories to the study of cities has yet to be realized.

By nature, working on the complex constructs of the risk city and urban resilience requires "complex thinking and complex methods" (see de Roo and Juotsiniemi 2010: 90). The complexity approach offers a suitable method of generating the kind of insights we seek with regard to the future trajectories of cities. It also forces us to adopt a more holistic view (Batty 2007).

2.3 The Risk City and the Dilemma of Lack

In practice, the risk city can be considered to be in a state of lack, to use the terminology proposed by French psychoanalyst and thinker Jacques Lacan. Because it does not engage in practices to address all types of risk, many aspects of the trust perceptions among city residents go unsatisfied. In this sense, the risk city seeks to provide what people feel they are lacking. I concur with Gunder and Hillier, who introduce the concept of "lack" to spatial planning theories and maintain that "we tend to welcome anything that gives us new positive identity and belief...We especially welcome new concepts and ideas that can somehow give us a sense of control or 'certainty' over the contingent complexities of life, including our environment and our future" (Gunder and Hillier 2009: 29). Accordingly, the practices and plans of the risk city are believed to reduce doubt and uncertainty and

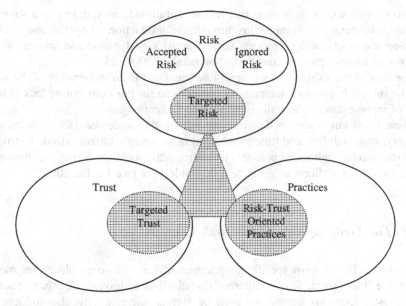

Fig. 2.2 The lack of the risk city

promise certainty in the future. However, these 'imaginary' "plans and their pre-scribed solutions lack" (Gunder and Hillier 2009: 29). The risk city, which lives upon unstable foundations, asks us to "continue to plan for certainty, even if we know—in our heart—that it is merely illusion and rationalization" (Gunder and Hillier 2009: 29).

The risk city does not tackle all type of risk (see Fig. 2.2). In addition to the *targeted risk* that it seeks to address and to mitigate there is also *accepted risk*, which the risk city accepts and agrees to live with without challenging (for various political, economic, and cultural reasons), and *ignored risk*, which the risk city consciously or subconsciously disregards. This has a direct impact on regions of trust in the city, as some are constructed and reinforced through specific practices and plans (both real and 'imaginary') while others go untended as a function of the risk city's decision to accept some risks and ignores others. The shaded areas in Fig. 2.2 represent targeted risk and trust and their induced practices. The non-shaded areas represent the risk city's "lack": that is, its unaddressed realms of risk and trust and the corresponding practices in the city that either never come into existence or are never employed.

The concept of "constitutive lack" initially emerged as an ontological concept in the work of Jacques Lacan (Robinson 2005). The basic claim of Lacanian theory is that identity, whether individual or social, is founded on a lack that has social and political consequences because it rules out the possibility of achieving substantial improvements in any area on which this fundamental negativity bears. In this way, it is "ineradicable" (Mouffe 2000). Because of its particular lack—that is, the regions of trust within its borders that remain unfulfilled, the risk city as a subject will always attempt to compensate through the production of spatial and social planning and practices. This process keeps the risk city dynamic and innovative in the eyes of its residents. According to Stavrakakis (2007: 25).

The idea of the subject as lack cannot be separated from recognition of the fact that the subject is always attempting to compensate for this constituting lack at the level of representation, through continuous identification acts.

There is an emptiness to the risk city left by the perceived lack of security, certainty, sustainability, and trust by which it is necessarily characterized. It strives to fill this gap through social practices and "pragmatic social construction," through utopian vision and efforts to generate a desirable state (see Laclau 2003).

2.3.1 The Geographies of the Risk City

The risk city has its own specific geographies of fear. As some places are more vulnerable than others, fear is differentially distributed throughout urban spaces. Like coastal locations during tsunamis or storms, the risk city also influences peoples' spatial practices and spatial behaviors. For example, urban residents usually prefer to walk along less risky routes, to avoid familiar sites of crime and violence, and to bypass the enclaves of "others." As urban crises result in feelings of mistrust between groups, people look to alternative social institutions for spaces of trust (Jabareen 2006). In an abstract way, *spaces of risk* represents the fear of not feeling 'at home' and the risk of losing all that is familiar and ordinary, all we have grown culturally and socially accustomed to and take for granted, and all that is unique to us as a collective group, as a community, and, indeed, as a nation.

2.4 The Risk City as a Risking City

Throughout this book, I use the term "risking city" to refer to modern cities' contribution to human risk in general and to their own risk in particular. From this perspective, the risk city is also a *risking city* due to its modes of production and consumption. Since the Industrial Revolution and the globalization of development, cities have emerged as major producers of gas emissions. Through their production and consumption of energy and materials, they pose a very real risk to themselves and to humanity in its entirety. Some cities, primarily in developed countries, have been promoting environmental agendas, plans, public policies, and actions aimed at lessening the emission of their cities, although these efforts have been limited in their scope and impact. The vast majority of developing countries, however, have failed to exhibit such environmental concerns for an array of reasons stemming from available resources, political structure, knowledge, and high levels of poverty.

2.5 Conclusions

The risk city is a conceptual construct of risk, trust, and practices that sheds light on the contemporary conditions of the risks and uncertainties facing cities and their residents, and the actions that are undertaken (or not undertaken) to cope with them. At the same time, it is a praxis that links theory with practice. The risk city acts to acquire knowledge regarding future uncertainties related to environmental and climate change and other significant risks and to construct socio-political and spatial frameworks aimed at responding to and countering these uncertainties. The risk city dynamically mobilizes its various resources in an effort to determine its own future. In this way, it makes use of the conditions of risk in a positive manner to creatively reconstruct itself and to address people, energy, and environmental, spatial, and economic development. Lack is one of its primary driving forces, as it seeks to win the trust of its residents, but can be only partially successful at doing so. After all, like its constituent components of risk and trust, the risk city is socially and culturally constructed and means different things to different people in different social and political contexts.

References

Alesina, A., & La Ferrara, E. (2005). Ethnic diversity and economic performance. *Journal of Economic Literature, 43*, 762–800.

Arrow, K. J. (1972). Gifts and exchanges. *Philosophy and Public Affairs, 1*, 343–362.

Beck, U. (1992). *Risk society: Towards a new modernity*. London: Sage.

Beck, U. (2005). *Power in the global age*. Cambridge: Polity Press.

Blumenfeld, H. (1969). Criteria for judging the quality of the urban environment. In H. J. Schmandt & W. Bloomberg (Eds.), *The quality of urban life 3* (pp. 137–163). California: Sage Publications.

Barber, B. (1983). *The logic and limits of trust*. New Brunswick, NJ: Rutgers University Press.

Batty, M. (2007). *Complexity in city systems: Understanding, evolution, and design*. MA: MIT Press.

Beamish, T. (2001). Environmental hazard and institutional betrayal: Lay-public perceptions of risk in the San Luis Obispo County oil spill. *Organization and Environment, 14*(1), 5–33.

Beck, U. (1996). Risk society and the provident state. In S. Lash, B. Szerszynski, & B. Wynne (Eds.), *Risk, environment and modernity*. London: Sage.

Beck, U., Giddens, A., & Lash, S. (1994). *Reflexive modernization: Politics, tradition, and aesthetics in modern social order*. Cambridge, UK: Polity.

Bickerstaff, K., Simmons, P., & Pidgeon, N. (2008). Constructing responsibilities for risk: Negotiating citizen-state relationships. *Environment and Planning A, 40*, 1312–1330.

Bowater, D. (2010). Climate change lies are exposed. *Sunday Express*. http://www.express.co.uk/news/uk/196642/Climate-change-lies-are-exposed. Accessed August 31, 2010.

Cebulla, A. (2000). Trusting community developers: The influence of the form and origin of community groups on residents' support in Northern Ireland. *Community Development Journal, 35*(2), 109–119.

Coleman, J. S. (1988). Social capital in the creation of human capital. *American Journal of Sociology, 94*, 95–120 (Issue Supplement: Organizations and Institutions: Sociological and Economic Approaches to the Analysis of Social Structure).

Cook, J., & Wall, T. (1980). New work measures of trust, organizational commitment, and personal need nonfulfillment. *Journal of Occupational Psychology, 53*, 39–52.

Crouch, E. A. C., & Wilson, R. (1982). *Risk/benefit analysis*. Cambridge, UK: Ballinger.

Cvetkovich, G., & Winter, P. L. (2003). Trust and social representations of the management of threatened and endangered species. *Environment and Behavior, 35*, 286–307.

De Roo, G., & Juotsiniemi, A. (2010). Planning and complexity. In *Book of Abstracts: 24th AESOP Annual Conference* (p. 90). Finland.

Deleuze, G., & F. Guattari. (1991). *What Is Philosophy?* New York: Columbia University Press.

Dhesi, A. S. (2000). Social capital and community development. *Community Development Journal, 35*(3), 199–214.

Douglas, M. A., & Wildavsky, A. (1982). *Risk and culture: An essay on the selection of technological and environmental dangers*. Berkeley, CA: University of California Press.

Dunn, J. (1984). The concept of trust in the politics of John Locke. In R. Rorty, J. B. Schneewind, & Q. Skinner (Eds.), *Philosophy in history: Essays on the historiography of philosophy* (pp. 279–301). Cambridge: Cambridge University Press.

Earle, T. C., & Cvetkovich, G. T. (1995). *Social Trust: Towards a cosmopolitan society*. New York: Praeger.

Erikson, K. (1994). *A new species of trouble: The human experience of modern disaster*. New York: Norton.

Fukuyama, F. (1995). *Trust: The social virtues and the creation of prosperity*. New York: Free Press.

Gambetta, D. (1988). Can we trust trust? In D. Gambetta (Ed.), *Trust: Making and breaking cooperation relations* (pp. 213–237). New York: Basil Blackwell.

Giddens, A. (1976). *The rules of sociological method*. London: Hutchinson.

Giddens, A. (1990). *The consequences of modernity*. Stanford, California: Stanford University Press.

Giddens, A. (1991). *Modernity and self-identity. Self and society in the late modern age*. Cambridge: Polity Press.

Giddens, A. (1999). *Runaway world: Risk*. Hong Kong: Reith Lectures.

Greenberg, M., & Schneider, D. (1997). Neighborhood quality, environmental hazards, personality traits, and resident actions. *Risk Analysis 17*(2), 169–175.

Guinnane, T. W. (2005). *Trust: A concept too many*. Economic Growth Center, Yale University. http://www.econ.yale.edu/growth_pdf/cdp907.pdf.

Gunder, M., & Hillier, J. (2009). *Planning in ten words or less: A Lacanian entanglement with spatial planning*. Farnham: Ashgate.

Guseva, A., & Rona-Tas, A. (2001). Uncertainty, risk, and trust: Russian and American credit card markets compared. *American Sociological Review, 66*(5), 623–646.

Häkli, J. (2009). Geographies of rust. In J. Häkli & C. Minca (Eds.), *Social capital and urban networks of trust* (pp. 13–35). Farnham: Ashgate.

Harvey, D. (1996). *Justice, nature and the geography of difference*. Cambridge, MA: Blackwell.

Healy, S. (2004). A "post-foundational" interpretation of risk: Risk as "performance". *Journal of Risk Research, 7*(3), 227–296.

Heimer, C. (1985). *Reactive risk and relative risk: Managing moral hazard in insurance contracts*. Berkeley: University of California Press.

Hillier, J. (2010). Introduction. In J. Hillier & P. Healey (Eds.), *The ashgate research companion to planning theory: Conceptual challenges for spatial planning* (pp. 1–34). Farnham: Ashgate.

Hoogenboom, M., & Ossewaarde, R. (2005). From iron cage to pigeon house: The birth of reflexive authority. *Organizational Studies, 26*(4), 601–619.

IAC—InterAcademy Council. (2011). InterAcademy council report recommends fundamental reform of IPCC management structure. http://reviewipcc.interacademycouncil.net.

IPCC—Intergovernmental Panel on Climate Change. (2007). *Climate change 2007: Fourth assessment report of the intergovernmental panel on climate change*. Cambridge, MA: Cambridge University Press.

Jabareen, Y. (2006a). Sustainable urban forms: Their typologies, models, and concepts. *Journal of Planning Education and Research, 26*(1), 38–52.

Jabareen, Y. (2006b). Conceptualizing space of risk: The contribution of planning policies to conflicts in cities—lessons from Nazareth. *Planning Theory and Practice, 7*(3), 305–323.

Jabareen, Y., & Carmon, N. (2010). Community of trust: A socio-cultural approach for community planning and the case of Gaza. *Habitat International, 34*(4), 446–453.

Jarvis, D. (2007). Risk, globalization and the state: A critical appraisal of Ulrich Beck and the world risk society thesis. *Global Society, 27*(1), 23–46.

Jasanoff, S. (1986). *Risk management and political culture: A comparative study of science in the policy context.* New York: Russell Sage Foundation.

Jasanoff, S. (1999). The songlines of risk. *Environmental Values, 8*(2), 135–152.

Josang, A., & Lo Presti, S. (2004). Analysing the relationship between risk and trust. In: *Second International Conference on Trust Management* (iTrust 2004) (pp. 135–145). Oxford, UK. March 29–April 1, 2004.

Kelley, H., & Thibaut, J. (1978). *Interpersonal relations: A theory of interdependence.* New York: Wiley.

Knack, S., & Keefer, P. (1997). Does social capital have an economic payoff? A cross-country investigation. *Quarterly Journal of Economics, 112*, 1251–1288.

Kollok, P. (1994). The emergence of exchange structures: An experimental study of uncertainty, commitment and trust. *American Journal of Sociology, 100*, 313–345.

Kumar, A., & Paddison, R. (2000). Trust and collaborative planning theory: The case of the Scottish planning system. *International Planning Studies, 5*(2), 205–223.

Laclau, E. (2003). Why do empty signifiers matter to politicians? In S. Zizek (Ed.), *Jacques lacan* (Vol. III, pp. 305–313). London: Routledge.

Leigh, A. (2006). Trust, inequality and ethnic heterogeneity. *Economic Record, 82*(258), 268–280.

Lidskog, R., Soneryd, L., & Uggla, Y. (2006). Making transboudary risks governable: Reducing complexity, constructing spatial identity, and ascribing capabilities. *A Journal of the Human Environment, 40*(2), 111–120.

Locke, J. (1976). In E. S. De Beer (Ed.), *The correspondence of John Locke* (Vol. 5). Oxford: Clarendon Press.

Luhmann, N. (1979). *Trust and power.* New York: Wiley.

Marc, B., & de Roo, G. (2010). What can spatial planning learn from ecological management? Exploring potentials of panarchy for spatial management. In *Book of Abstracts: 24th AESOP Annual Conference* (pp. 93–94). Finland.

McGlade, J., & Garnsey, E. (2006). The nature of complexity. In E. Garnsey & J. McGlade (Eds.), *Complexity and co-evolution* (pp. 1–21). MA: Edward Elgar.

Molm, L., Takahashi, N., & Peterson, G. (2000). Risk and trust in social exchange: An experimental test of a classical proposition. *American Journal of Sociology, 105*(5), 1396–1427.

Mouffe, C. (2000). *The democratic paradox.* London: Verso.

Nature (Ed.). (2010). Cities: The century of the city. *Nature, 467*, 900–901.

November, V. (2008). Spatiality of risk. *Environment and Planning A, 40*(7), 1523–1527.

Ollman, B. (1990). Putting dialectics to work: The process of abstraction in Marx's method. *Rethinking Marxism, 3*(1), 26–74.

Petak, W. J., & Atkisson, A. A. (1982). *Natural hazard risk assessment and public policy.* New York: Springer.

Portugali, J. (2010). *Complexity, cognition and the city.* Berlin: Springer.

Putnam, R. (1995). Bowling alone: America's declining social capital. *Journal of Democracy, 6*(1), 65–78.

Renn, O., & Rohrmann, B. (Eds.). (2000). *Cross-cultural risk perception research.* Dordrecht: Kluwer.

Robinson, A. (2005). The political theory of constitutive lack: A critique. *Theory and Event, 8*(1).

Rohrmann, B. (2006). *Cross-cultural comparison of risk perceptions: Research, results, relevance.* Presented at the ACERA/SRA Conference. http://www.acera.unimelb.edu.au.

Ross, C., Mirowsky, J., & Pribesh, S. (2001). Powerlessness and the amplification of threat: Neighborhood disadvantage, disorder, and mistrust. *American Sociological Review, 66*(4), 568–591.

Rosenzweig, C., Solecki, W. D., Hammer, S. A., & Mehrotra, S. (2011). *Climate change and cities: First assessment report of the urban climate change research network.* Cambridge: Cambridge University Press.

Sampson, R., & Groves, B. (1989). Community structure and crime: Testing social-disorganization theory. *American Journal of Sociology, 94*(4), 774–802.

Satterthwaite, D. (2007). *The transition to a predominantly urban world and its underpinnings. Human Settlements Series: Urban Change-4.* http://www.iied.org/pubs/pdfs/10550IIED.pdf.

Schindler, P. L., & Thomas, C. C. (1993). The structure of interpersonal trust in the workplace. *Psychological Reports, 73*, 563–573.

Schmidt, W., & Posner, B. (1982). *Managerial values and expectations: The silent power in personal and organizational life.* New York: American Management Association.

Shaw, C., & McKay, H. (1942). *Juvenile delinquency and urban areas.* Chicago: University of Chicago Press.

Skogan, W. (1990). *Disorder and Decline.* Berkeley, CA: University of California Press.

Snary, C. (2004). Understanding risk: The planning officers' perspective. *Urban Studies, 41*(1), 33–55.

Sommerfield, J., Kouyate, M. S., & Sauerborn, R. (2002). Perceptions of risk, vulnerability, and disease prevention in rural Burkina Faso: Implications for community-based health care and insurance. *Human Organization, 2*, 139–146.

Stavrakakis, Y. (2007). *The lacanian left.* Albany: SUNY.

Storper, M. & Manville, M. (2006). Behaviour, preferences and cities: Urban theory and urban resurgence. *Urban Studies, 43*(8), 1247–1274.

Taylor, R. B., & Shumaker, S. A. (1990). Local Crime as a natural hazard: Implications for understanding the relationship between disorder and fear of crime. *American Journal of Community Psychology, 18*, 619–641.

The World Bank. (2011). *Guide to climate change adaptation in cities.* The International Bank for Reconstruction and Development—The World Bank.

UNISDR-International Strategy for Disaster Reduction. (2010). *Making cities resilient: My city is getting ready.* 2010–2011 World Disaster Reduction Campaign.

Urry, J. (2005). The complexity turn. *Theory, Culture and Society, 22*(5), 567–582.

Uslaner, E. (2002). *The moral foundations of trust.* Cambridge: Cambridge University Press.

Chapter 3
Planning Practices for Cities Countering Climate Change

3.1 Introduction

Climate change poses new risks and uncertainties which often lie outside our range of experience and have the potential to affect the social, economic, ecological, and physical systems of any given city (IPCC 2007: 719). In this way, climate change and its resulting uncertainties challenge the practices and concepts of planning, creating a need to rethink and revise current planning methods. When it comes to cities, "we are still at the stage of setting agendas and directions for research, and there are more questions than answers" (Priemus and Rietveld 2009: 425). At this point, as Harriet (2010: 20) points out, "we simply do not know what the impact of many of the initiatives that have been undertaken over the past two decades has been or what these achievements might amount to collectively."

The recent literature on the subject shows clearly that climate change is already happening. This is vividly reflected in the American Northeast, one of the world's most built-up environments which is home to approximately 64 million people, including the city of New York, and which presents a compelling case of dramatic climate change that is already having an impact (Horton et al. 2014). This region is tremendously vulnerable to the impact of climate change, as was clearly demonstrated when superstorms Irene and Sandy struck in 2011 and 2012, respectively. In the past 115 years (between 1895 and 2011), temperatures in the Northeast have increased by almost 2 °F (0.16 °F per decade), and precipitation has increased by approximately five inches (0.4 in. per decade), or more than 10 % (Kunkel et al. 2013). The increase in precipitation in the Northeast has been more extreme than in any other region in the United States, reflected in the over 70 % increase in the amount of precipitation falling in very heavy events between 1958 and 2010 (Horton et al. 2014; Groisman et al. 2013). Coastal flooding has also increased due to a rise in sea level of approximately 1 foot since 1900—an increase that exceeds the global average of approximately 8 in. (Church et al. 2010).

© Springer Science+Business Media Dordrecht 2015
Y. Jabareen, *The Risk City*, Lecture Notes in Energy 29,
DOI 10.1007/978-94-017-9768-9_3

Although countering climate change in cities is a complex and multidisciplinary undertaking that demands a "paradigm shift" toward transdisciplinary thinking (Bosher 2008; CCC 2010; Coaffee and Bosher 2008; Dainty and Bosher 2008; Godschalk 2003), most of the literature on the subject is still fragmented and fractional in scope and typically overlooks the multidisciplinary nature of the issue. Without a doubt, focusing on one or a small number of aspects of the problem ultimately results in partial or inaccurate conclusions, misrepresentation of the multiple causes of climate change impact, ineffective policy, and "unfortunate and sometimes disastrous unintended consequences" (Bettencourt and Geoffrey 2010). For this reason, the major theoretical challenge regarding planning for countering climate change in the urban context appears to be the development of a multidisciplinary framework, *a praxis* integrating the theories and the practices used by the risk city, to cope with the outcomes of climate change.

The following section outlines the grounded theory methods for building the conceptual framework of Planning for Countering Climate Change (PCCC) in the urban context. The chapter's third section demonstrates the conceptual framework and its concepts, and the final section offers a number of conclusions.

3.2 Methodology: How to Construct a Conceptual Framework

In my theorization of the *Conceptual Framework for Countering Climate Change* (CFCCC), I employ the ontological conceptualization of "planes of immanence" and the term "concept" used by Deleuze and Guattari (1991). A conceptual framework is a plane of immanence and a network of interlinked concepts that together provide a comprehensive understanding of a phenomenon (Jabareen 2009: 51). It is an "object of construction" (Bonta and Protevi 2004, 62–63) defined by what it contains as well as the interwoven relations between its concepts. A conceptual framework, therefore, is not merely a collection of concepts but rather a construct composed of "consistent" concepts in which each plays an integral role and is intrinsically linked to the others. This enables it to better provide "not a causal/analytical setting but, rather, an *interpretative approach* to social reality" and to our understanding of the multiple and interlinked concepts it encompasses (Jabareen 2009: 51).

According to Deleuze and Guattari (1991: 15), "every concept has components and is defined by them" and "there is no concept with only one component." Components define the consistency of the concept and are distinct and heterogeneous, yet also inseparable from one another. It is a multiplicity, but not every multiplicity defines a concept. Every concept must be understood "relative to its own components, to other concepts, to the plane on which it is defined, and to the problem it is supposed to resolve" (Deleuze and Guattari 1991: 21). Moreover, every concept has its own history and typically contains "bits" or components originating in other concepts. In other words, all concepts relate back to other concepts; they are always created from something, and

Fig. 3.1 The process of building a framework for cities countering climate change

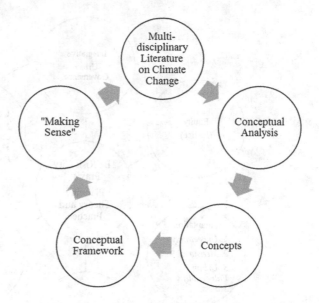

cannot be created from nothing. To build the CFCCC, I employed the method of conceptual analysis (Jabareen 2009)—a grounded theory technique that aims to generate, identify, abstract, and trace the major concepts that together constitute the theoretical framework of the phenomenon (Jabareen 2009). This methodology delineates the following stages in conceptual framework building: (a) mapping selected data sources; (b) surveying literature and categorizing selected data; (c) identifying and naming the concepts; (d) deconstructing and categorizing the concepts; (e) integrating the concepts; (f) synthesis, resynthesis, and making it all make sense; (g) validating the conceptual framework; and (h) rethinking the conceptual framework.

The primary source of data is multidisciplinary literature and multiple bodies of knowledge and states of affairs. The texts selected for conceptual framework analysis inclusively represent the relevant spatial, ecological, environmental, economic, social, cultural, and political multidisciplinary literatures that focuses on the phenomenon under study. Data is taken from a variety of source types, including books, articles, newspapers, essays, and interviews. Most of the texts and much of the data is representative of theories belonging to specific disciplines. When we adopt a multidisciplinary approach, however, these discipline-oriented theories become the empirical data of the conceptual framework for analysis (Fig. 3.1).

3.3 The Concepts of the Risk City Praxis

The conceptual framework of evaluation is composed of eight concepts of assessment, all of which are directed in one way or another toward climate change adaptation and GHG reduction, as represented in Fig. 3.2.

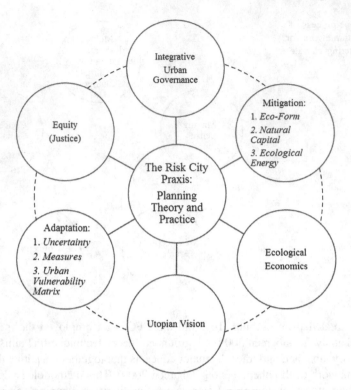

Fig. 3.2 Concepts of planning countering climate change (PCCC)

This praxis, or the theoretical-practical framework of the risk city, is directed toward reducing the risk and uncertainties stemming from climate change. The framework consists of six concepts, each of which belongs to different contents and theoretical settings. Although the concepts of mitigation and adaptation are dominant in the literature on environmental and ecological climate change, these criteria or concepts are theoretically and practically insufficient for providing a full account of the praxis of the risk city from the perspective of climate change. Other concepts make a significant contribution to our understanding of the theoretical and practical settings of the risk city. For example, the concept of equity represents the elements of justice and ethicality that are related to climate change oriented practices, whereas integrative urban governance suggests how to manage the risk city through integration among institutions and the building of new organizational capacities to meet the challenge of the risk of climate change. Integrating ecological economics, or the "green economy," into the practices of the risk city helps facilitate and promote the existence of the rest of practices. Finally, the vision of these practices—the vision of the risk city—reframes the problem and the lack of the current situation and calls for filling this lack and the existing gaps in the present or in the future.

The concepts of the risk city praxis are as follows:

3.3.1 Utopian Vision

The Utopian vision of the risk city calls for filling the lack of the current social and spatial conditions of the city and the existing "gaps" in the present or in the future. The risk city always maintains dreams and visions of how to contend with its unstable conditions and its lack. David Harvey (2000: 195) suggests that utopian visions and dreams "never fade away" and "are omnipresent in the hidden signifiers of our desires." Friedmann (2002: 3) posits that utopian thinking is about imagining

> a future that is radically different from what we know to be the prevailing order of things...
> a way of breaking through the barriers of convention into a sphere of the imagination where
> many things beyond our everyday experience become possible. We need a constructive
> imagination to help us create the fictive worlds of our dreams, of dreams worth struggling for.

This concept is concerned with a plan's future vision. Usually, urban planning seeks to bring about a different and more desirable future. Theoretically, the power of visionary or utopian thinking lies in its inherent ability to envision the future in terms of radically new forms and values (de Geus 1999). An urban vision incorporating climate change as a central theme is of the utmost importance to practitioners, decision makers, and the public. Visionary frames are important in climate change, as they serve to identify problematic conditions and the need for change, to propose future alternatives, and to urge all stakeholders to act in concert to affect change. Climate change planning visions must provide people with an interpretive framework that enables them to better understand how the issue is related to their own lives in the present and future, and to the world at large (Taylor 2000; Benford and Snow 2000: 614). Recently, many countries, states, cities, and communities appear to have adapted a vision in which climate change plays a central role (e.g. CCC 2010; NYC 2009). The utopian vision concept evaluates a plan's visionary and utopian aspects regarding future urban life and the city's potential role in climate change mitigation.

The significance of visioning and utopias lie in their role in considering potential urban futures and motivating critical perspectives on cities and social change (Pinder 2010: 345). Convincingly, David Pinder argues that "utopias are not simply of historical interest but also matter—indeed, are necessary—for current pressing discussions about how more humanizing urban spaces and ways of living might be struggled for and produced." Pinder suggests that even if there are dark sides to utopias in urban planning, utopian thinking is important and still part of the human spirit.

3.3.2 Equity

Justice, equality, fairness, and equity are significant moral terms of the recent literature on cities and planning. Useful literature has also been produced on "the right to the city," which has constituted a leading concept for framing rights and addressing injustices and inequalities in contemporary cities (i.e. Amin and Thrift 2002; Bernner et al. 2012; Friedmann 2002; Harvey and Potter 2009; Elden 2004; Beauregard and Bounds 2000; Brodie 2000; Fainstein 2009; Marcuse 2009). Theoretically, the frameworks of "the right to city" and "the just city" are based on moral choices and a normative political approach that incorporates rights based on a "moral" claim "founded on fundamental principles of justice" (Marcuse 2012: 35). These two theoretical frameworks are based on a normative agenda aimed at the planning of urban space in the hope of achieving a more "harmonious, and just urban form" (Harvey and Potter 2009: 40) and at creating "a revitalized, cosmopolitan, just, and democratic city" (Fainstein 2009: 20). In the words of Friedmann (2002: 104), these frameworks constitute a "concrete imagery of utopian thinking to propose steps that would bring us a little closer to a more just world."

Justice's role in the risk city is based on two basic assumptions: the assumption that space production should be based on justice and on a moral agenda, and the assumption that the more just the city is, the more efficiently it will cope with climate change.

Equity encompasses issues of social and environmental justice and fairness and therefore plays a central role in evaluating climate change policies. The concept of equity is used to evaluate a plan's social aspects, such as environmental justice, public participation, and methods of addressing each community's vulnerability to climate change (urban vulnerability matrix).The impact of climate change and climate change mitigation policies is unevenly distributed and "socially differentiated," and is therefore a matter of local and international distributional equity and justice (Adger 2001: 929; Bruce et al. 1996; Davies et al. 2008; IPCC 2007; Kasperson and Kasperson 2001; O'Brien et al. 2004; Paavola and Adger 2006; Tearfund 2008).

Mohai et al. (2009) go further, holding that climate change actually increases social inequality and that adaptive and resilience resources have apparently been distributed unequally, as demonstrated by the aftermath of Hurricane Katrina in the United States. Climate change injustice in countries around the world occurs along ethnic, gender, class, and racial lines (Mohai et al. 2009; Adger et al. 2006) and even emerges among neighborhoods and communities. Some hold that the reverse is also true: that is, that true inequality leads to greater environmental degradation and that a more equitable distribution of power and resources would result in improved environmental quality (Boyce et al. 1999; Agyeman et al. 2002; Solow 1991; Stymne and Jackson 2000). Moreover, the communities that are most vulnerable to climate change impact are typically those who live in more vulnerable, high-risk locations and may lack skills and adequate infrastructure and services (Satterthwaite 2008).

All societies contain individuals and groups who are more vulnerable than others and lack the capacity to adapt to climate change (Schneider et al. 2007: 719). In the context of climate change, vulnerability refers to the "degree to which a system is susceptible to, and unable to cope with, adverse effects of climate change, including climate variability and extremes. Vulnerability is a function of a system's exposure, its sensitivity, and its adaptive capacity" (CCC 2010: 61). A society's vulnerability is influenced by its development path, physical exposure, resource distribution, social networks, government institutions, and technological development (IPCC 2007: 719–720).

3.3.3 Mitigation

The risk city applies mitigation measures to help reduce the risk posed to the city. Mitigation practices and policies also contribute to the global effort to counter climate change. On the whole, the climate change literature proposes two major types of measures for countering climate change: mitigation measures, which aim at reducing greenhouse gas emissions (GHG), and adaptation measures, which are geared toward contending with the unavoidable effects (CEC 2009). The debate's reduction to these two kinds of measures misses the point and dismisses many other significant kinds of measures employed as part of the practices of the risk city. The risk city, therefore, requires more than these two kinds of policies.

Mitigation refers to an "action to reduce the sources (or enhance the sinks) of factors causing climate change, such as greenhouse gases" (CCC 2010: 61), and to "the reduction of GHG emissions and their capture and storage in order to limit the extent of climate change" (Bulkeley 2010: 2.2). The EU has recently agreed to reduce emissions to 20 % below 1990 levels by 2020. However, the agreement could potentially be amended to deliver a 30 % reduction if undertaken as part of an international agreement in which other developed countries agree to comparable reductions and appropriate contributions by more economically advanced developing countries.

Mitigation measures includes several measures borrowed from various bodies of knowledge (particularly spatial planning, resources, and energy), as outlined below:

3.3.3.1 Natural Capital

Natural capital refers to "the stock of all environmental and natural resource assets, from oil in the ground to the quality of soil and groundwater, from the stock of fish in the ocean to the capacity of the globe to recycle and absorb carbon" (Pearce et al. 1990: 1). Maintaining constant natural capital is an important criterion for sustainability (Pearce and Turner 1990, 44; Neumayer 2001; Geldrop and Withagen 2000). The stock of natural capital should not decrease, as this could endanger the ecological system and threaten the ability of future generations to generate wealth and maintain their well-being. Natural capital provides us with themes of the consumption and—equally as important—the renewal of natural assets that are used for development, such as land, water, air, and open spaces.

3.3.3.2 Ecological Energy

Energy is a "defining issue of our time" (Yumkella 2009: 1) and "access to clean and affordable energy is one of the main prerequisites for sustainable economic and social development" (UNIDO 2009: 6). Ecological energy is perhaps the most important concept for climate change. The clean, renewable, and efficient use of energy is a central theme in all planning for the achievement of climate change objectives. This concept evaluates the manner in which a plan addresses the energy sector and whether it proposes strategies to reduce energy consumption and to use new alternative and cleaner energy sources. Ecological energy suggests that energy should be based on new low-carbon technologies in order to meet the target of emissions reduction. For example, the target for 2050 in the UK is to reduce emissions by 80 % relative to 1990 levels. Low-carbon technologies will be vital in generating cleaner forms of electricity, which can then be used for electric vehicles, heating, and more energy efficient buildings (CCC 2010).

3.3.3.3 Eco-Form: The Urban Form of the Risk City

The urban form and its typologies are significant to the risk city. The risk city has a desired form and typologies that help it cope with threats and uncertainties. In past centuries, the fortifications of the walled city provided its residents with a sense of trust and reduction of risk, and the city's risk-oriented typologies were reflected in walls, entrances, hidden passages, and the like. The present-day risk city seeks alternative forms and typologies to mitigate and cope with uncertainty and risk.

City form matters when it comes to countering climate change and coping with risk. Elsewhere, I discuss the importance of city form for achieving sustainability (Jabareen 2006). However, with the growing awareness of risk and climate change, ideas of sustainability and practices regarding urban form appear to be in need of theoretical and practical updating to include not only adaptation measures but the concepts of risk reduction and contributing to the mitigation agenda. This would better enable the urban forms and their design concepts to contribute to planning for the mission of countering climate change and reducing risk.

City form means "the spatial pattern of the large, inert, permanent physical objects in a city" (Lynch 1981: 47). Form is a result of aggregations of more or less repetitive elements and the bringing together of many different elements and concepts into an urban pattern. It is a composite of characteristics related to land use patterns, the transportation system, and urban design (Handy 1996: 152–153). Urban patterns are made up largely of a limited number of relatively undifferentiated types of recurring elements that combine with one another. Hence, these patterns have strong similarities and can be grouped conceptually into what are called concepts (Lozano 1990: 55). Specifically, elements of concepts might be street patterns, block size and form, street design, typical lot configuration, the layout of parks and public spaces, etc. (Jabareen 2006).

The question of city form is crucial for achieving sustainability on the urban level as well as for coping with risk. The emergence of the concept of sustainability has revived discussion about the form of cities (see: Jabareen 2004, 2006). Undoubtedly, the urban form of the risk city should be re-conceptualized as a concept not only of sustainability but also of risk, mitigation, and adaptation. The form of the contemporary city itself has been perceived as a source of environmental problems (Alberti 2003; Beatley and Manning 1997; EPA 2001; Haughton 1999: 69; Hildebrand 1999: 16; Jabareen 2006; Newman and Kenworthy 1989). In *Our Built and Natural Environments* (EPA 2001), the U.S. Environmental Protection Agency (EPA) concludes that the urban form directly affects habitat, ecosystems, endangered species, and water quality through land consumption, habitat fragmentation, and replacement of natural cover with impervious surfaces. In addition, urban form affects travel behavior which, in turn, affects air quality; the premature loss of farmland, wetlands, and open space; soil pollution and contamination; global climate; and noise pollution (Cervero 1998: 43–48).

Since the rise of the industrial revolution, cities, their production activities, and their transportation modes have been a major source of climate change and the global environmental crisis. The spatial typologies of metropolitan areas, cities, towns, villages, neighborhoods, and houses have also had a considerable impact on local sustainability and global climate change. Urban form has a powerful environmental impact that affects the consumption of energy, land, and nature; travel behavior; greenhouse gas emissions; and water and soil pollution. Urgent changes are therefore needed not only in our production and consumption behaviors but also in the design of our cities' forms and all other human habitats. Ultimately, forms of human habitat, which reflect many aspects of our modes of production and consumption, are highly important for achieving not only improved environmental conditions but also safety and security. Elsewhere, I have suggested a theoretical model for assessing the sustainability of urban forms (Jabareen 2006).

An important, unanswered question regarding the adaptation-mitigation dilemma is: what is the most desirable urban form of the risk city? For example, what is optimal density and compactness for the purpose of reducing risk and saving energy? The physical form of a city affects its habitats and ecosystems, the everyday activities and spatial practices of its inhabitants, and, ultimately, climate change. This component represents the spatial planning, architecture, design, and the ecologically-desired form of the city and its components (such as buildings and neighborhoods). On this basis, I propose assessing and structuring the urban form of the risk city in accordance with the following three concepts:

1. *Multiple Responsibility*—By virtue of its function in the context of the safety and the protection of inhabitants from disasters and threats, urban form plays an important role in the risk city. It should therefore be oriented toward enhancing safety and promoting a sense of trust, and not simply the feeling of community advocated by "new urbanism" and "neo-traditionalism."
2. *Adaptability*—Urban form should not be fixed or stagnated but rather flexible and capable of adapting in the face of uncertainty. Understanding future

vulnerability, exposure, and the response capacity of interlinked human and natural systems is challenging due to the number of interacting social, spatial, city form, economic, and cultural factors that have been incompletely considered to date (see: IPCC 2014).The concept of adaptability here is about adjusting the city form at the present to adapt urban spaces to anticipated and uncertain harm. Adaptation means "controlling uncertainty—either by taking action now to secure the future or by preparing actions to be taken in case an event occurs" (Abbott 2005: 237). In this sense, the responsibility and adaptability of the urban form are critical for the city. Without a doubt, the risk stemming from climate change and its resulting uncertainties challenge our conventional perspectives on urban forms. Adaptability, therefore, suggests that planning should be uncertainty-oriented rather than based on the conventional planning approaches. From this perspective, one essential requirement for countering climate change is uncertainty management that includes adaptation policies. In order to effectively contend with the new challenges of climate change to our contemporary cities, planners must develop a greater awareness for adaptation at the spatial level and on the level of form. We ultimately expect that form adjustments might enhance resilience and reduce vulnerability to expected climate change impact at the focus of the planning process (Adger et al. 2007: 720).

3. *Sustainability*—This concept represents the mitigation measures and functions of urban form that exist in addition to social dimensions such as the creation of a sense of community and the like. Elsewhere (Jabareen 2006), I have suggested the following set of nine planning typologies, or criteria of evaluation, which are helpful in understanding the impact of different features of urban form on the practices of risk in general and mitigation policies in particular:

   (a) **Compactness** refers to urban contiguity and connectivity and suggests that future urban development should take place adjacent to existing urban structures (Wheeler 2002). Compact urban space can minimize the need to transport energy, materials, products, and people (Elkin et al. 1991). Intensification, a major strategy for achieving compactness, uses urban land more efficiently by increasing the density of development and activity, and involves developing previously undeveloped urban land, redeveloping existing buildings and previously developed sites, subdivisions and conversions, and additions and extensions (Jenks 2000: 243).

   (b) **Sustainable Transport** suggests that planning should promote sustainable modes of transportation by means of traffic reduction, trip reduction, the encouragement of non-motorized travel (such as walking and cycling), transit-oriented development, safety, equitable access, and renewable energy sources (Cervero 1998; Clercq and Bertolini 2003).

   (c) **Density** is the ratio of people or dwelling units to land area. Density affects climate change through differences in the consumption of energy, materials, and land for housing, transportation, and urban infrastructure. High density planning can save significant amounts of energy (Carl 2000; Walker and Rees 1997; Newman and Kenworthy 1989).

(d) **Mixed Land Uses** refers to the diversity of functional land uses, such as residential, commercial, industrial, institutional, and transportation. It allows planners to locate compatible land uses close to one another in order to decrease the distance of travel between activities. This encourages walking and cycling and reduces the need for car travel by ensuring that employment, shopping, and leisure facilities are all located in close proximity (Parker 1994; Alberti 2000; Van and Senior 2000; Thorne and Filmer-Sankey 2003).

(e) **Diversity** is "a multidimensional phenomenon" that promotes other desirable urban features, such as a larger variety of housing types, building densities, household sizes, ages, cultures, and incomes (Turner and Murray 2001: 320). Diversity is vital for cities, as without it the urban system declines as a living place (Jacobs 1961) and the resulting homogeneity of built forms, which often produces unattractive monotonous urban land-scapes, leads to increased segregation, car travel, congestion, and air pol-lution (Wheeler 2002).

(f) **Passive Solar Design** aims to reduce energy demands and to facilitate the best use of passive energy through specific planning and design measures, such as orientation, layout, landscaping, building design, urban materials, surface finish, vegetation, and bodies of water. This facilitates the optimum use of solar gain and microclimatic conditions and reduces the need to heat and cool buildings using conventional energy sources (Owens 1992; Thomas 2003; Yannis 1998: 43).

(g) **Greening**, or bringing "nature into the city," makes positive contributions to many aspects of the urban environment, including biodiversity, the lived-in urban environment, urban climate, economic attractiveness, community pride, and health and education (Beatley 2000; Swanwick et al. 2003; Forman 2002; Dumreicher et al. 2000; Beer et al. 2003; Ulrich 1999).

(h) **Renewal and Utilization** refers to the process of reclaiming the many sites that are no longer appropriate for their original intended use and that can be used for a new purpose, such as brownfields. Cleaning, rezoning, and developing contaminated sites are key aspects of revitalizing cities and neighborhoods and contributing to their sustainability and to a healthier urban environment.

(i) **Planning Scale** both influences and is influenced by climate change. For this reason, a desirable planning scale should be considered and integrated into plans on the regional, municipal, district, neighborhood, street, site, and building levels. Planning that moves from the macro level to the micro level has a more holistic and positive impact on climate change.

Some critical contradictions and conflicts exist between aspects of urban form and planning concepts, most notably between sustainability and mitigation on the one hand, and adaptation measures on the other hand. Hamin and Gurran (2009: 293) convincingly contend that "despite clear recognition by the IPCC of the need to ensure that adaptation actions do not undermine mitigation attempts, let alone broader sustainability goals (IPCC 2007), surprisingly little research exists on the types of conflicts that might arise in practice." My proposed framework for the urban form of the risk city contains a conflict of which we should remain aware and treat accordingly: on the one hand, sustainability and mitigation planning measures suggest compactness, more dense and compact urban spaces, mass transportation, and time-space-energy reduction; on the other hand, compact and dense urban spaces may prove more dangerous when under a threat or during an extreme event.

What is the optimal density and desired compactness for sustainability and mitigation and for adaptation and safety under extreme circumstances? I suggest that we begin by studying the case of New York City, which underwent extreme experiences during superstorms Sandy and Irene, and by considering the manner in which density affects resident behavior and evacuation and the resilience of high-density, moderate-density, and low-density areas during these storms.

Hamin and Gurran (2009: 241) contend that a key element of adaptation is the fact that many actions require leaving more land as open space and a less dense built environment. Moreover, adding (or not removing) space-using greenery is an important step in preventing or treating urban heat island effects (Stone 2008). Buildings that are more moderate in height and placed to enable ventilation between individual dwellings provide adaptation to higher temperatures, but tend to reduce density. While there is little adaptation benefit from low density sprawling development, under adaptation it appears that moderate density with significant fingers of green infrastructure running through the city may be the most effective form (Hamin and Gurran 2009: 241).

3.3.4 Integrative Urban Governance

To attempt to contend with the complex risks and impact of climate change at the city level, the risk city needs good integrative governance. To enhance the urban governance of the risk city, the assumption is that we need to expand and improve local capacity by increasing knowledge, providing resources, establishing new institutions, enhancing good governance, and increasing local autonomy (Allman et al. 2004; Bai 2007; Corfee-Morlot 2009; Harriet 2010; Holgate 2007; Lankao 2007; Bulkeley et al. 2009; Kern 2008: 56).

Urban policies for countering climate change are crucial factors in bringing the governance of global environmental problems to urban contexts (Betsill and Bulkeley 2007; Bulkeley and Newell 2010; Biermann and Pattberg 2008; Harriet 2010; Okereke et al. 2009). Local authorities in all countries play a critical role in mitigating and adapting to climate change (Satterthwaite 2008). Indeed, as Harriet (2010) points out, "such institutional barriers do not operate within a political vacuum, and more often than not, it is the urban political economies of climate change that matter most in enabling and constraining effective action."

Because of the great uncertainty characterizing the context in which it is undertaken, urban policies and planning to counter climate change is too complex for conventional approaches to accommodate. Adaptation measures of "hard infrastructure" must be complemented by "soft infrastructure" and other resilience measures, such as, improving institutional coordination, public communication, and rapid decision making abilities in order to improve the ability to recover from the catastrophic effects of natural disasters (Rodin and Rohaytn 2013: 7). This context poses new challenges for collaboration among public, private, and civil institutions and organizations on all levels. Integrating the many different stakeholders and agents into the planning process is essential for achieving climate change objectives. The "ability of a governance system to adapt to uncertain and unpredicted conditions is a new notion" (Mirfenderesk and Corkill 2009: 152). Adaptive management requires new planning strategies and procedures that transcend conventional planning approaches by integrating uncertainties into the planning process and prioritizing stakeholders' expectations in an uncertain environment. Plans should also be "flexible enough to quickly adapt to our rapidly changing environment" (Mirfenderesk and Corkill 2009).

The concept of integrative urban governance evaluates the integrative framework of city planning and adaptive management under conditions of uncertainty and the spectrum of collaboration that a plan proposes. Such governance is an essential element of resilience, which may also be understood as "diversity and redundancy in our systems and rewiring their interconnections" in a manner that "enables their functioning even when individual parts fail" (NYS 2100 Commission 2013: 7).

3.3.5 Ecological Economics

This concept evaluates the ecological economic aspects of a plan, including the economic engines it puts into place in order to meet climate change objectives. The idea is to create opportunities to integrate climate change planning, protection, and development approaches into the city's economic development decisions and strategies, and for these elements to dictate reforms in the areas of investment and insurance and risk management related to natural disasters and other emergencies (NYS 2100 2013: 10).

Ecological economics is based on the assumption that environmentally sound economics can play a decisive role in achieving climate change objectives in a capitalist world. Cities that are committed to climate change mitigation and sustainability should stimulate markets for green products and services, promote environmentally friendly consumption, and contribute to urban economic development by creating a cleaner environment (Hsu 2006: 11). In this spirit, the American Recovery and Reinvestment Plan proposed by President Barack Obama calls for spurring "job creation while making long-term investments in energy and infrastructure," and increasing "production of alternative energy" (White House 2009).

The report issued by New York State following Hurricane Sandy posits that "building a 21st century resilience strategy comes with significant economic opportunities. Newly conceived infrastructure investments will be rooted in rebuilding smarter while also creating the jobs of tomorrow, including green jobs" (NYS 2100 2013: 7). The European Commission (EC 2010: 9) has argued that a "well designed climate policy, including a sufficiently high and predictable carbon price, could contribute to foster technological change, innovation in green activities and green growth." The American Recovery and Re-investment Act, with an estimated total cost of $787 billion, allocated approximately $68 billion to clean energy investment, including smart grids, energy efficiency in buildings, local and state renewable energy and energy efficiency efforts, and R&D for energy storage, as well as significant investment for carbon capture and storage (CCS). The Council of Economic Advisors estimates that clean energy investments will create more than 700,000 job-years of employment by the end of 2012" (Executive Office of the President 2010). In addition, a recent EC study (2010) estimates the 2007 global market for green technologies at a volume of €1400 billion, and energy efficiency and environmentally friendly energy is projected to reach a volume of €1645 billion by 2020 (Federal Ministry for the Environment et al. 2009). Moreover, according to UNEP/NEF (2009), between 2002 and 2008, global annual public and private new investment in renewable energy and energy efficiency increased from $7.1 to $118.9 billion (UNEP 2005).

3.3.6 Adaptation

Without question, the risk city needs adaptation practices to contend with damage when it actually occurs, and a crucial element of countering climate change must therefore be uncertainty management that includes adaptation policies. In the context of climate change, adaptation is defined as an "adjustment of behavior to limit harm, or exploit beneficial opportunities, arising from climate change" (CCC 2010: 60). Most cities and countries appear to be applying mitigation policies to address the human causes of climate change by reducing greenhouse gas emissions but have failed to apply adaptation policies as well. According to the CCC (2010), "even with strong international action on mitigation, past and present emissions mean that the climate will continue to change." In this way, adaptation and mitigation are not alternatives but rather two complementary approaches to the problem, both of which are necessary.

Adaptation can be thought of as encompassing the following three main components:

3.3.6.1 Uncertainty

Uncertainty can be defined as "a perceived lack of knowledge, by an individual or group, which is relevant to the purpose or action being undertaken and its outcomes" (Abbott 2009: 503). Today more than ever, we need to acknowledge that environmental uncertainties pose new challenges to our cities and their communities, and challenge the way we have been thinking about their management and planning. As vulnerability assessments often ignore the non-climatic drivers of future risk (Storch and Downes 2011) and uncertainties, it is important to carry out as extensive a mapping as possible of the scenarios of uncertainties that may affect our cities. In the context of contemporary cities making efforts to develop a greater awareness of the need for policies that might eventually enhance resilience and reduce vulnerability to expected climate change impacts (Adger et al. 2001; Vellinga et al. 2009), uncertainty has a critical impact on urban vulnerability and requires the assessment of environmental risks and hazards that are difficult to predict but must be taken into account in city planning and risk management.

The new urban uncertainties posed by climate change challenge the concepts, procedures, and scope of planning. To cope with the new challenges, planners must develop a greater awareness and place mitigation and policies for adaptation, or actual adjustments that might eventually enhance resilience and reduce vulnerability to expected climate changes, at the focus of the planning process (Adger et al. 2007: 720).

3.3.6.2 Measures

Planners must also develop a better understanding of the risks that climate change poses to infrastructure, households, and communities. To address these risks, planners have two types of uncertainty or adaptation management at their disposal: (1) ex-ante management, or actions taken to reduce and/or prevent risky events; and (2) ex-post management, or actions taken to recover losses after a risky event (Heltberg et al. 2009).

When we adopt adaptation measures, we acknowledge that the climate will continue to change and that we must take measures to reduce the risks (Priemus and Rietveld 2009). From this perspective, adaptation to climate change must be considered indispensable (Vellinga et al. 2009).

The Commission of the European Communities (CEC) (2009) suggests that "even if the world succeeds in limiting and then reducing GHG emissions, our planet will take time to recover from the greenhouse gases already in the atmosphere. Thus we will be faced with the impact of climate change for at least the next 50 years. We need therefore to take measures to adapt." The CEC (2009) regrets the "piecemeal manner" in which adaptation policies have been implemented and concludes that "a more strategic approach is needed to ensure that timely and effective adaptation measures are taken, ensuring coherency across different sectors and levels of governance." Plans should also be "flexible enough to quickly adapt to our rapidly changing environment" (Mirfenderesk and Corkill 2009).

3.3.6.3 Urban Vulnerability Matrix

This component is of critical significance for the resilient city and its contribution to the spatial and socio-economic mapping of future risks and vulnerabilities.

The purpose of the Vulnerability Analysis Matrix is to analyze and identify types, demography, intensity, scope, and spatial distribution of environmental risk, natural disasters, and future uncertainties in cities. In addition, this concept seeks to address how hazards, risks, and uncertainties affect various urban communities and urban groups.

In the context of climate change, vulnerability refers to the "degree to which a system is susceptible to, and unable to cope with, adverse effects of climate change, including climate variability and extremes. Vulnerability is a function of a system's exposure, its sensitivity, and its adaptive capacity" (CCC 2010: 61). The Vulnerability Analysis Matrix is composed of the following four main components, which together determine its scope and its environmental, social, and spatial nature:

Demography of Vulnerability This component assesses and examines the demographic and socio-economic aspects of urban vulnerability. It assumes that there are individuals and groups within all societies who are more vulnerable than others and lack the capacity to adapt to climate change (Schneider et al. 2007: 719). Demographic, health-related, and socio-economic variables affect the ability of individuals and urban communities to face and cope with environmental risk and future uncertainties by impacting risk mitigation, response, and recovery in the event of natural disasters (Blaikie et al. 1994; Ojerio et al. 2010). The vulnerability of individuals and communities is shaped by many different factors, the most important of which are income, education and language skills, gender, age, physical and mental capacity, accessibility to resources and political power, and social capital (Cutter et al. 2003; Morrow 1999; Ojerio et al. 2010; United Nations Division for the Advancement of Women 2001). As a result, socio-economically weak communities are more vulnerable to negative impacts such as property loss, physical harm, and psychological distress (Ojerio et al. 2010; Fothergill and Peek 2004).

Informality This concept assesses the scale and social, economic, and environmental conditions of informal urban spaces. Informal spaces are unplanned, chaotic, and disorderly (Roy 2010) and the scale and human condition of informal places within a city are assumed to have a significant impact on its vulnerability. According to UN-HABITAT (2014), much urban expansion in developing cities takes place outside the official and legal frameworks of building codes, land use regulations, and land transactions. Resilience requires the inclusion of poor vulnerable communities and informal places throughout the city and the greater metropolitan area. Informal spaces are more likely to be vulnerable than others because of their low-income population and lack of infrastructure and services. Moreover, because of their socio-spatial character and large populations, contemporary cities are more vulnerable to a variety of risks and have the potential to become generators of new risks, such as failed infrastructure and services, environmental urban degradation, and the expansion of informal settlements. These aspects make many urban inhabitants more vulnerable to natural hazards and risks (UNISDR 2010).

The Spatial Distribution of Vulnerability This component assesses the spatial distribution of risks, uncertainties, vulnerability, and vulnerable communities in cities. Environmental risks and hazards are not always evenly distributed geographically, and some communities may be affected more than others. For example, those who are close to the shore may be affected more harshly by tsunamis than others. Mapping the spatial distribution of risks and hazards is critical for planning and management at the present and for the future. In addition, the communities that are most vulnerable to climate change impacts are usually those who live within more vulnerable, high-risk locations that may lack skills, adequate infrastructure, and services (Satterthwaite 2008).

3.4 Conclusions: The Planning for Countering Climate Change Framework

This chapter has theorized the practices of the risk city and proposed Planning for Countering Climate Change (PCCC) as a framework for enabling our contemporary risk cities to counter climate change in the best manner possible. As we have seen, PCCC has a number of critical characteristics:

1. PCCC differs from conventional planning methods in its approach, data analysis, visioning, and procedures. Otherwise, it would not be able to effectively assume responsibility for the lives of millions of people.
2. The planning practices of PCCC are informed by the concepts of risk and trust.
3. PCCC acknowledges and addresses the complexities of countering climate change at the urban level. Cities are complex socio-spatial and environmental phenomena, and therefore capturing their complexity requires multidisciplinary methods.
4. At its core, PCCC is based not only on demographic, economic, and spatial analysis but also on the analysis of risk and uncertainties. Knowledge regarding climate change impact has become a critical resource for spatial planning. PCCC Identifies human spaces, places, and assets which are vulnerable to extreme weather events, storm surges, sea-level rises, temperature changes, seismic events, etc.
5. In addition to the analysis of the existing adaptation measures on the city and neighborhood level, PCCC employs the Urban Vulnerability Matrix to more thoroughly understand the social-spatial distribution of risk and uncertainties. In this context, it operates to address the threats posed at the specific level of communities and social groups.
6. Public involvement and participation is of critical significance for PCCC, which uses GIS and other tools to support analysis, visualization, public participation, and decision-making.

7. PCCC creates opportunities to integrate planning, protection and development approaches into economic development decisions and strategies.

8. Unlike traditional planning approaches, PCCC incorporates adaptation measures. In other words, it is based on the premise that planning must also think in terms of urban defensibility or protection of the city, and therefore consider the state of critical infrastructures, and also in terms of providing protection through new measures such as natural infrastructure projects and coastal ecosystem restoration to create additional lines of storm defenses. PCCC also focuses on protecting critical systems, such as transit system and tunnels against severe flooding, strengthening vulnerable main transportation routes and the like. In PCCC, future risk and uncertainties contribute to spatial planning and, subsequently, to the location of new developments and growth patterns that avoid areas with high vulnerability. PCCC seeks to ensure alternative functional routes and infrastructures in the event of an extreme event.

9. Unlike conventional planning approaches, PCCC incorporates energy as a major guiding concept in planning cities and communities.

10. PCCC approaches land use differently than conventional planning, based on the analysis of data related to risk and uncertainties.

11. PCCC offers a dynamic and flexible conceptual framework and methodology for addressing uncertainties and is capable of accommodating new emerging risks that may affect cities in a dramatic way, integrating them into a framework for countering climate change.

12. PCCC integrates the development of scenario-planning capability as part of its procedures and outcomes. This scenario-planning helps explore policy options regarding where to build, what to build, and how to strengthen communities in the areas of greatest risk.

13. PCCC involves people and communities in the planning, from visioning to scenario planning aimed at informing and guiding decisions "about long-term rebuilding efforts, future investment plans, and the level to which we rely upon 'soft' solutions or harden and upgrade our infrastructure" (NYS 2100 2013: 12).

14. PCCC proposes systematic and rigorous ways of comparing the manner in which different cities counter climate change.

15. Due to its "easy to grasp" nature, PCCC has the potential to facilitate greater awareness among scholars professionals, decision makers, and the public as a whole regarding the current and future direction of cities in the arena of climate change issues. The proposed multifaceted conceptual framework can help us determine what needs to be done to increase the resilience of our cities, thereby enabling us to work more effectively toward making them more safe and secure.

References

Abbott, J. (2005). Understanding and managing the unknown: The nature of uncertainty in planning. *Journal of Planning Education and Research, 24*(2), 237–251.

Abbott, J. (2009). Planning for complex metropolitan regions: A better future or a more certain one? *Journal of Planning Education and Research, 28*, 503–517.

Adger, W. N. (2001). Scales of governance and environmental justice for adaptation and mitigation of climate change. *Journal of International Development, 13*(7), 921–931.

Adger, W. N., Paavola, J., Huq, S., & Mace, M. J. (2006). *Fairness in adaptation to climate change*. Cambridge, MA: MIT Press.

Adger, W. N., Eakin, H., & Winkels, A. (2007). Nested and networked vulnerabilities in South East Asia. In L. Lebel et al. (Eds.), *Global environmental change and the south-east Asian region: An assessment of the state of the science*. Washington, D.C.: Island Press.

Agyeman, J., Bullard, R. D., & Evans, B. (2002). Exploring the nexus: Bringing together sustainability, environmental justice and equity. *Space and Polity, 6*(1), 77–90.

Alberti, M. (2000). Urban form and ecosystem dynamics: Empirical evidence and practical implications. In K. Williams, E. Burton, & M. Jenks (Eds.), *Achieving sustainable urban form* (pp. 84–96). London: E & FN Spon.

Allman, L., Fleming, P., & Wallace, A. (2004). The progress of english and welsh local authorities in addressing climate change. *Local Environment, 9*, 271–283.

Alberti, M., Marzluff, J., Shulenberger, E., Bradley, G., Ryan, C., & Zumbrunnen, C. (2003). Integrating humans into ecology: Opportunities and challenges for urban ecology. *BioScience 53*(12), 1169.

Amin, A., & Thrift, N. (2002). *Cities: Reimagining the urban*. Cambridge: Polity Press.

Bai, X. (2007). Integrating global environmental concerns into urban management: The scale and readiness arguments. *Journal of Industrial Ecology, 11*, 15–29.

Beatley, T. (2000). *Green urbanism: Learning from European cities*. Washington, D.C.: Island Press.

Beatley, Timothy, & Kristy, Manning. (1997). *Ecology of place: Planning for environment, economy, and community*. Washington, D.C.: Island Press.

Beauregard, R., & Bounds, A. (2000). Urban citizenship. In E. F. Isin (Ed.), *Democracy, citizenship and the global city* (pp. 243–256). New York: Routledge.

Beer, A., Delshammar, T., & Schildwacht, P. (2003). A changing understanding of the role of greenspace in high-density housing: A European perspective. *Built Environment, 29*(2), 132–143.

Benford, R. D., & Snow, D. A. (2000). Framing processes and social movements: An overview and assessment. *Annual Review of Sociology, 26*, 611–639.

Betsill, M. M., & Bulkeley, H. (2007). Looking back and thinking ahead: A decade of cities and climate change research. *Local Environment, 12*, 447–456.

Bettencourt, L., & West, G. (2010) A unified theory of urban living. *Nature, 467*(7318), 912–913.

Biermann, F., & Pattberg, P. (2008). Global environmental governance: taking stock, moving forward. *Annual Review of Environment and Resources, 33*, 277–294.

Blaikie, P., Cannon, T., Davis, I., & Wisner, B. (1994). *At risk: Natural hazards, people's vulnerability, and disasters*. London: Routledge.

Bonta, M., & Protevi, J. (2004). *Deleuze and geophilosophy: A guide and glossary*. Edinburgh: Edinburgh University Press.

Bosher, L. S. (2008). *Hazards and the built environment: Attaining built-in resilience*. London: Taylor & Francis.

Boyce, J. K., Klemer, A. R., Templet, P. H., & Willis, C. E. (1999). Power distribution, the environment, and public health: A state-level analysis. *Ecological Economics, 29*(1), 127–140.

Brenner, N., Marcuse, P., & Mayer, M. (Eds.). (2012). *Cities for people, not for profit: critical urban theory and the right to the city*. New York: Routledge.

Brodie, J. (2000). Imagining democratic urban citizenship. In E. Isin (Ed.), *Democracy, citizenship and the global city* (pp. 110–128). New York: Routledge.

Bruce, J. P., Lee, H., & Haites, E. F. (Eds.). (1996). *Climate change 1995: Economic and social dimensions of climate change: Contribution of working group III to the second assessment report of the intergovernmental panel on climate change.* Cambridge: Cambridge University Press.

Bulkeley, H., & Newell, P. (2010). *Governing climate change.* London: Routledge.

Bulkeley, H., Schroeder, H., Janda, K., Zhao, J., Armstrong, A., Chu, S. Y., & Ghosh, S. (2009). *Cities and climate change: The role of institutions, governance and urban planning.* Paper presented at the World Bank 5th Urban Symposium on Climate Change, June, Marseille.

Carl, P. (2000). Urban density and block metabolism. In K. Steemers & S. Yannas (Eds.), *Proceedings of PLEA 2000 Architecture, City, Environment* (pp. 343–347). London: James & James.

CCC—Committee on Climate Change Adaptation. (2010). *How Well Prepared is the UK for Climate Change?* www.theccc.org.uk.

CEC—The Commission of the European Communities. (2009). *White paper: Adapting to climate change: Towards a European framework for action.* Brussels.

Cervero, R. (1998). *The transit metropolis: A global inquiry.* Washington, D.C.: Island Press.

Church, J. A., Woodworth, P. L., Aarup, T., & Wilson, W. S. (2010). *Understanding sea-level rise and variability.* New York: Wiley.

Clercq, F., & Bertolini, L. (2003). Achieving sustainable accessibility: An evaluation of policy measures in the Amsterdam area. *Built Environment, 29*(1), 36–47.

Coaffee, J., & Bosher, L. (2008). Integrating counter-terrorist resilience into sustainability. *Proceeding of the Institute of Civil Engineering: Urban Design and Planning, 161*(DP2), 75–84.

Corfee-Morlot, J., Kamal-Chaoui, L., Donovan, M. G., Cochran, I., Robert, A., & Teasdale, P. J. (2009). *Cities, climate change and multilevel governance.* OECD Environmental Working Papers 14: 2009, OECD publishing.

Cutter, S. L., Boruff, B. J., & Shirley, W. L. (2003). Social vulnerability to environmental hazards. *Social Science Quarterly, 84*(2), 242–261.

Dainty, A. R. J., & Bosher, L. S. (2008). Afterword: Integrating resilience into construction practice. In L. S. Bosher (Ed.), *Hazards and the built environment: Attaining built-in resilience.* London: Taylor & Francis.

Davies, M., Guenther, B., Leavy, J., Mitchell, T., & Tanner, T. (2008). *Climate change adaptation, disaster risk reduction and social protection: Complementary Roles in agriculture and rural growth?* Institute of Development Studies Centre for Social Protection and Climate Change and Disasters Group. IDS: Institute of Developing Studies. http://www.climategovernance.org/docs/SP-CC-DRR_idsDFID_08final.pdf.

de Geus, M. (1999). *Ecological Utopias: Envisioning the sustainable society.* International Books.

Deleuze, G., & Guattari, F. (1991). *What is philosophy?.* New York: Columbia University Press.

Dumreicher, H., Levine, R. S., & Yanarella, Ernest J. (2000). The appropriate scale for "low energy": Theory and practice at the Westbahnhof. In K. Steemers & S. Yannas (Eds.), *Proceedings of PLEA Architecture, City, Environment 2000* (pp. 359–363). London: James & James.

EC-The European Commission. (2010). *Commission staff working document.* Brussels, 26.5.2010. http://ec.europa.eu/environment/climat/pdf/26-05-2010working_doc.pdf.

Elden S. (2004). *Understanding Henri Lefebvre: Theory and the Possible.* New York: Continuum.

Elkin, T., McLauren, D., & Hillman, M. (1991). *Reviving the city: Towards sustainable urban development, policy studies.* London: Institute/Friends of the Earth.

EPA—United States Environmental Protection Agency. (2001). *Our built and natural environments: A technical review of the interactions between land use, transportation, and environmental quality.* EPA 231-R-01–002. Online http://www.smartgrowth.org/.

Executive Office of the President, Council of Economic Advisers (2010). The Economic Impact of the American Recovery and Reinvestment Act of 2009, Second Quarterly Report. January 13, 2010. http://www.whitehouse.gov/the-press-office/economic-impact-american-recovery-and-reinvestment-act-2009-second-quarterly-report.

Fainstein, S. (2009). Planning and the just city. In P. Marcuse, J. Connolly, J. Novy, I. Olivo, C. Potter, & J. Steil (Eds.), *Searching for the just city: Debate in urban theory and practice* (pp. 19–39). New York: Routledge.

FME—Federal Ministry for the Environment, Nature Conservation and Nuclear Safety. (2009). *GreenTech made in Germany 2.0.* http://www.bmu.de/files/pdfs/allgemein/application/pdf/greentech2009_en.pdf.

Forman, R. T. (2002). The missing catalyst: Design and planning with ecology. In B. T. Johnson & K. Hill (Eds.), *Ecology and design: Frameworks for learning*. Washington, DC: Island Press.

Fothergill, A., & Peek, L. (2004). Poverty and disasters in the United States: A review of recent sociological findings. *Natural Hazards, 32*(1), 89–110.

Friedmann, J. (2002). *The prospect of cities*. Minneapolis, MN: University of Minnesota Press.

Geldrop, J., & Withagen, C. (2000). Natural capital and sustainability. *Ecological Economics, 32* (3), 445–455.

Godschalk, D. R. (2003). Urban hazards mitigation: Creating resilient cities. *Natural Hazards Review, 4*(3), 136–143.

Groisman, P. Y., Knight, R. W., & Zolina, O. G. (2013). Recent trends in regional and global intense precipitation patterns. In R. A. Pielke Sr. (Ed.), *Climate vulnerability* (pp. 25–55). Massachusetts: Academic Press.

Hamin, E. M., & Gurran, N. (2009). Urban form and climate change: Balancing adaptation and mitigation in the U.S. and Australia. *Habitat International, 33*(3), 238–246.

Handy, S. (1996). Methodologies for exploring the link between urban form and travel behavior. *Transportation Research: Transport and Environment: D, 2*(2), 151–165.

Harriet, B. (2010). Cities and the governing of climate change. *Annual Review of Environment and Resources, 35*, 2.1–2.25.

Harvey, D. (2000). *Space of Hope*. Edinburgh: Edinburgh University Press.

Harvey, D., & Potter, C. (2009). The right to the just city. In P. Marcuse, J. Connolly, J. Novy, I. Olivo, C. Potter, & J. Steil (Eds.), *Searching for the just city: Debate in urban theory and practice* (pp. 40–51). New York: Routledge.

Haughton, Graham. (1999). Environmental justice and the sustainable city. In D. Satterthwaite (Ed.), *Sustainable cities*. London: Earthscan.

Heltberg, R., Siegel, P. B., & Jorgensen, S. L. (2009). Addressing human vulnerability to climate change: Toward a 'no-regrets' approach. *Global Environmental Change, 19*(2009), 89–99.

Hildebrand, F. (1999). *Designing the city: Towards a more sustainable urban form*. London: E & FN Spon.

Holgate, C. (2007). Factors and actors in climate change mitigation: A tale of two South African cities. *Local Environment, 12*, 471–484.

Horton, R., G. Yohe, W. Easterling, R. Kates, M. Ruth, E. Sussman, A. Whelchel, D. Wolfe, & Lipschultz, F. (2014). Ch. 16: Northeast. Climate change impacts in the United States: The third national climate assessment. In J. M. Melillo, T. C. Richmond, & G. W. Yohe (Eds.), *U.S. Global Change Research Program* (pp. 371–395). doi:10.7930/J0SF2T3P.

Hsu, D. (2006). *Sustainable New York city. New York City: Design trust for public space and the New York city office of environmental coordination*. New York City: New York City Office of Environmental Coordination.

IPCC—Intergovernmental Panel on Climate Change. (2007). *Climate change 2007: Fourth assessment report of the intergovernmental panel on climate change*. Cambridge, MA: Cambridge University Press.

IPCC—Intergovernmental Panel on Climate Change. (2014). *Climate change 2014: Impacts, adaptation, and vulnerability*. http://ipccwg2.gov/AR5/images/uploads/IPCC_WG2AR5_SPM_Approved.pdf.

Jabareen, Y. (2004). A knowledge map for describing variegated and conflict domains of sustainable development. *Journal of Environmental Planning and Management, 47*(4), 632–642.

Jabareen, Y. (2006). Sustainable urban forms: Their typologies, models, and concepts. *Journal of Planning Education and Research, 26*(1), 38–52.

Jabareen, Y. (2009). Building conceptual framework: Philosophy, definitions and procedure. *International Journal of Qualitative Methods, 8*(4), 49–62.

Jacobs, J. (1961). *The death and life of great American cities.* New York: Random House.

Jenks, M. (2000). The acceptability of urban intensification. In K. Williams, E. Burton, & M. Jenks (Eds.), *Achieving sustainable urban form.* London: E & FN Spon.

Kasperson, R. E., & Kasperson, J. X. (2001). *Climate change, vulnerability and social justice.* Stockholm: Stockholm Environment Institute.

Kern, K., & Alber, G. (2008). Governing climate change in cities: modes of urban climate governance in multi-level systems. In *OECD Conference Proceedings of Competitive Cities and Climate Change* (pp. 171–196). Paris, Milan, Italy: OECD. October 9–10, 2008. http://www.oecd.org/dataoecd/54/63/42545036.pdf.

Kunkel, K. E., Stevens, L. E., Stevens, S. E., Sun, L., Janssen, E., Wuebbles, D., Rennells, J., DeGaetano, A., Dobson, J. G. (2013). *Regional climate trends and scenarios for the U.S. national Climate assessment: Part 1.*

Lozano, E. E. (1990). *Community design and the culture of cities: The crossroad and the wall.* Cambridge: Cambridge University Press.

Lynch, Kevin. (1981). *A theory of good city form.* Cambridge: The MIT Press.

Marcuse, P. (2009). From justice planning to commons planning. In P. Marcuse, J. Connolly, J. Novy, I. Olivo, C. Potter, & J. Steil (Eds.), *Searching for the just city: Debate in urban theory and practice* (pp. 91–102). New York: Routledge.

Marcuse, P. (2012). Whose right(s), to what city? In N. Brenner, P. Marcuse, & M. Mayer (Eds.), *Cities for people, not for profit: Critical urban theory and the right to the city* (pp. 24–41). New York: Routledge.

Mirfenderesk, H., & Corkill, D. (2009). Sustainable management of risks associated with climate change. *International Journal of Climate Change Strategies and Management, 1*(2), 146–159.

Mohai, P., Pellow, D., & Roberts, J. T. (2009). Environmental justice. *Annual Review of Environment and Resources, 34*, 405–430.

Morrow, B. H. (1999). Identifying and mapping community vulnerability. *Disasters, 23*(1), 1–18.

Neumayer, E. (2001). Do countries fail to raise environmental standards? An evaluation of policy options addressing regulatory chill. *International Journal of Sustainable Development, Inderscience Enterprises Ltd, 4*(3), 231–244.

Newman, P., & Kenworthy, J. (1989). Gasoline consumption and cities: A comparison of US cities with a global survey. *Journal of the American Planning Association, 55*, 23–37.

NYC-The City of New York, Mayor Michael R. Bloomberg. (2009). *PlaNYC: Progress Report 2009.*

NYS. (2013). *NYS2100 commission: Recommendations to improve the strength and resilience of the empire state's infrastructure.*

O'Brien, K., Leichenko, R., Kelkar, U., Venema, H., Aandahl, G., Tompkins, H., et al. (2004). Mapping vulnerability to multiple stressors: Climate change and globalization in India. *Global Environmental Change, 14*, 303–313.

Ojerio, R., Moseley, C., Lynn, K., & Bania, N. (2010). Limited involvement of socially vulnerable populations in federal programs to mitigate wildfire risk in Arizona. *Natural Hazards Review, 12*(1), 28–36.

Okereke, C., Bulkeley, H., & Schroeder, H. (2009). Conceptualizing climate governance beyond the international regime. *Global Environmental Politics, 9*, 58–78.

Owens, S. (1992). Energy, environmental sustainability and land-use planning. In B. Michael (Ed.), *Sustainable development and urban form* (pp. 79–105). London: Pion.

Paavola, J., & Adger, W. N. (2006). Fair adaptation to climate change. *Ecological Economics, 56* (4), 594–609.

Parker, T. (1994). *The land use—air quality linkage: How land use and transportation affect air quality*. Sacramento: California Air Resources Board.

Pearce, D., & Turner, R. K. (1990). *Economics of natural resources and the environment*. Baltimore: Johns Hopkins University Press.

Pearce, D., Edward, B., & Markandya, A. (1990). *Sustainable development: Economics and environment in the third world*. London: Earthscan Publications.

Pinder, D. (2010). Necessary dreaming: Uses of Utopia in urban planning. In J. Hillier & P. Healey (Eds.), *The ashgate research companion to planning theory: Conceptual challenges for spatial planning* (pp. 343–366).

Priemus, H., & Rietveld, P. (2009). Climate change, flood risk and spatial planning. *Built Environment, 35*(4), 425–431.

Rodin, J., & Rohaytn, F. G. (2013) *NYS 2100 commission: Recommendations to improve the strength and resilience of the empire state's infrastructure*. http://www.governor.ny.gov/sites/ governor.ny.gov/files/archive/assets/documents/NYS2100.pdf.

Romero Lankao, P. (2007). How do local governments in Mexico City manage global warming? *Local Environment, 12*, 519–535.

Roy, A. (2010). Informality and the politics of planning. In J. Hillier & P. Healey (Eds.), *Planning theory: Conceptual challenges for spatial planning* (pp. 87–107). Farnham: Ashgate Publishing.

Satterthwaite, D. (2008). *Climate change and urbanization: Effects and implications for urban governance*. Presented at UN Expert Group Meeting Population Distribution, Urbanization, Internal Migration Devopment. UN/POP/EGMURB/2008/16/.

Schneider, S. H., Semenov, S., Patwardhan, A., Burton, I., Magadza, C. H. D., Oppenheimer, M., et al. (2007). Assessing key vulnerabilities and the risk from climate change. Climate change 2007: Impacts, adaptation and vulnerability. In M. L. Parry, O. F. Canziani, J. P. Palutikof, P. J. van der Linden, & C. E. Hanson (Eds.), *Contribution of working group II to the fourth assessment report of the intergovernmental panel on climate change* (pp. 779–810). Cambridge, UK: Cambridge University Press.

Solow, R. (1991) *Sustainability: An economist's perspective*. The Eighteenth J. Seward Johnson Lecture. Woods Hole, MA: Woods Hole Oceanographic Institution.

Stone B Jr. (2008). Urban sprawl and air quality in large US cities. *Journal of Environmental Management, 86*(4), 688–698.

Storch, H., & Downes, N. K. (2011). A scenario-based approach to assess Ho Chi Minh City's urban development strategies against the impact of climate change. *Cities, 28*(6), 517–526.

Stymne, S., & Jackson, T. (2000). Intra-generational equity and sustainable welfare: A time series analysis for the UK and Sweden. *Ecological Economics, 33*(2), 219–236.

Swanwick, C., Dunnett, N., & Woolley, H. (2003). Nature, role and value of green space in towns and cities: An overview. *Built Environment, 29*(2), 94–106.

Taylor, D. E. (2000). The rise of the environmental justice paradigm: injustice framing and the social construction of environmental discourses. *American Behavioral Scientist, 43*, 508–580.

Tearfund. (2008). *Linking climate change adaptation and disaster risk reduction*. Web White Pap. http://www.tearfund.org/webdocs/Website/Campaigning/CCAandDRRweb.pdf.

Thomas, R. (2003). Building design. In R. Thomas & M. Fordham (Eds.), *Sustainable urban design: An environmental approach* (pp. 46–88). London: Spon Press.

Thorne, R., & Filmer-Sankey, W. (2003). Transportation. In T. Randall (Ed.), *Sustainable urban design* (pp. 25–32). London: Spon Press.

Turner, S. R. S., & Murray, M. S. (2001). Managing growth in a climate of urban diversity: South Florida's Eastward ho! Initiative. *Journal of Planning Education and Research, 20*, 308–328.

Ulrich, R. S. (1999). Effects of gardens on health outcomes: theory and research. In C. C. Marcus & B. Marni (Eds.), *Healing gardens: Therapeutic benefits and design recommendations*. New York: Wiley.

UN-Habitat—United Nations Human Settlements Programme (2014). The State of African Cities 2014: Re-imagining sustainable urban transitions. Nairobi: UN-Habitat.

UNEP—United Nations Environment Programme/New Economy Finance. (2009). Global trends in sustainable energy investment 2009. In *Analysis of trends and issues in the financing of renewable energy and energy efficiency*. UNEP.

UNIDO-United Nations Industrial Development Organization. (2009). *Energy and climate change: Greening the industrial agenda*. http://www.unido.org.

UNISDR—International Strategy for Disaster Reduction. (2010). *Making cities resilient: My city is getting ready*. 2010–2011 World Disaster Reduction Campaign.

United Nations Division for the Advancement of Women. (2001). *Environmental management and the mitigation of natural disasters: A gender perspective*. http://www.un.org/womenwatch/daw/csw/env_manage/documents/EGM-Turkey-final-report.pdf. July 7, 2009.

Van U. P., & Senior, M. (2000). The contribution of mixed land uses to sustainable travel in cities. In K. Williams, E. Burton & M. Jenks (Eds.), *Achieving sustainable urban form* (pp. 139–148). London: E & FN Spon.

Vellinga, P., Marinova, N. A., & van Loon-Steensma, J. M. (2009). Adaptation to climate change: A framework for analysis with examples from the Netherlands. *Built Environment, 35*(4), 452–470. .

Walker, L., & William, R. (1997). Urban density and ecological footprints—an analysis of Canadian households. In M. Roseland (Ed.), *Eco-city dimensions: Healthy communities, healthy planet*. Canada: New Society Publishers.

Wheeler, S. M. (2002). Constructing sustainable development/safeguarding our common future: Rethinnking sustainable development. *Journal of the American Planning Association, 68*(1), 110–111.

White House. (2009). http://www.whitehouse.gov/issues/Economy; see also http://www.recovery.gov/Pages/home.aspx.

Yannis, S. (1998). Living with the city: Urban design and environmental sustainability. In E. Maldonado & S. Yannas (Eds.), *Environmentally friendly cities* (pp. 41–48). London: James & James.

Yumkella, K. K. (2009). *Forward in energy and climate change: Greening the industrial agenda* (p. 1). UNIDO-United Nations Industrial Development Organization 2010. http://www.unido.org.

Chapter 4
Assessment Methods: Planning Practices Countering Climate Change

4.1 Introduction

Climate change is likely to affect the social, economic, ecological, and physical systems of any given city. Importantly, the impacts of climate change on urban systems depend not only on the level of emissions but also on how inherently vulnerable these systems are to the changing climate. The large uncertainties over future development and structure of cities, societies and economies mean that the assessment of climate change effects is complex (Hallegatte et al. 2011).

Thus, climate change and its resulting uncertainties challenge the conventional approaches to planning evaluation methods, creating a need to rethink and revise existing methods. Not surprisingly, various levels of government are currently struggling to find an assessment framework to help them assess their adaptation and mitigation policies (e.g. CCC 2010). Indeed, a planning literature review suggests a striking absence of such assessment methods of climate change oriented policies. Thus, this chapter aims to fill this urgent methodological gap and proposes a new multifaceted conceptual framework for assessing urban plans and planning policies aimed at coping with climate change in cities and communities.

In the urban context, there is a need to revise the existing conventional evaluation methods when it comes to assessing the impact of urban plans, public policies, and the preparedness of cities to meet the risks resulting from climate change. There are four primary reasons for such a revision, and for the exploration of new evaluation approaches. First, the multidisciplinary nature of *climate change* and its diverse effects on cities and their residents requires a multidisciplinary framework. A review of the planning literature reflects that, although current scholarship offers a number of criteria for assessing issues related to sustainability and climate change, it lacks a multifaceted evaluation framework for assessing plans' specific contributions to

© Springer Science+Business Media Dordrecht 2015
Y. Jabareen, *The Risk City*, Lecture Notes in Energy 29,
DOI 10.1007/978-94-017-9768-9_4

climate change mitigation. From this perspective, a more comprehensive assessment framework would be crucial for efforts to counter climate change.

Second, a major reason for the revision of existing planning evaluation methods is the uncertain and complex nature of the phenomenon of climate change. Crabbě and Leroy (2008: xi–xii) insightfully argue that the existing standard evaluation methods "might not be applicable to environmental policy-making given the complexities of the field." Furthermore, Mermet et al. (2010) conclude that environmental policies have been growing ever more complex and ambiguous as of late, sparking an urgent need for new evaluation approaches.

Third, the wide spectrum of people, institutions, and organizations that could be affected by climate change, and involved in countering it, suggests a need not for a complicated method, but rather for a simple and easy to comprehend evaluation framework that 'makes sense' and is easy to present to policy makers, practitioners, communities and the public at large.

Fourth, the existing environmental and planning evaluation methods mostly apply a set of assessment criteria, or indicators, which mostly appear as 'sustainable indicators'. They are frequently seen as sporadic and as not deriving from theoretical foundations or conceptual frameworks for climate change (Mermet et al. 2010). This paper argues that evaluation criteria that are not derived from a unified theory ultimately lack the theoretical and methodological consistency required to govern the evaluation method. Therefore, this paper proposes formulating theoretical foundations, or a conceptual framework for countering climate change *prior to* establishing evaluation criteria or indicators, and on that basis to develop the concepts of evaluation. In other words, the formulation of a theoretical or conceptual framework must be understood as a necessary precondition for the construction of an evaluation method.

To this end, this chapter proposes a theory-based evaluation method for understanding policies and programs that aim to counter the deterioration of climate change and to cope with its uncertainties and risks. In developing this framework, I was motivated by the need for an easy to grasp assessment method that allows planners, practitioners, policymakers, and interested members of the public at large to critically evaluate plans as they relate to the pressing issue of climate change. As climate change is a subject of multidisciplinary interest, the proposed framework draws on various bodies of knowledge.

The remainder of this chapter proceeds as follows. The first section reviews existing research on evaluation methods and climate change in general and in the planning discipline in particular. The second section describes the methodology of constructing a new conceptual framework for assessing planning policies aiming at countering climate change. The third section presents the Countering Climate Change Evaluation Method research results. The fourth section draws some conclusions for planners and scholars.

4.2 Planning Evaluation Methods and Climate Change

Evaluation is a multidisciplinary approach of the assessment of policies, plans, and programs (Baycan and Nijkamp 2005: 64). Ultimately, evaluation is aimed at understanding policies, their products, and their organizational and institutional context (Diez 2001: 41). It produces insight and recommendations aimed at improving policy design and generating new knowledge (Baycan and Nijkamp 2005; Diez 2001).

With the development of the social methodologies and methods of inquiry, the epistemological foundations of the evaluation methods have also changed over the years. Guba and Lincoln (1989: 22–49) identify four generations of evaluation methods. In the first, which they refer to as the *measurement generation* and which still exists today, the role of the evaluator was technical and instrumental. The second, or the *description generation*, was an objective-oriented approach that described patterns of strength and weaknesses with respect to stated objectives. The third generation, which they referred to as the *judgment generation*, aimed at reaching judgments of worth, and the evaluator assumed the role of judge. Fourth generation evaluation, or the *responsive constructivist generation*, provides an alternative constructivist inquiry paradigm in which the claims, concerns, and issues of stakeholders serve as organizational foci. Using a slightly different categorization, Vedung (2010) suggests four waves of evaluation diffusion: the *scientific wave*, based on "admittedly subjective goals"; the *dialogue-oriented wave*, based on discussions among stakeholders; the *neo-liberal wave*, which pushed for market orientation; and the *evidence-based wave*, which implies a renaissance for scientific experimentation.

In the field of urban planning, 'evaluation', or 'planning assessment', is an established field of research whose evolution has been closely tied to changes in planning theory and practice (Khakee et al. 2008). In recent years, evaluation of planning scenarios, policies and programs "has become a principle means of informing and supporting decision making" (Miller and Patassani 2005: xv). Moreover, the trends impacting the evolution of evaluation methods have also found expression in the planning field. The two major planning paradigms— rational planning and communicative planning—suggest differing evaluation methods. The evaluation methods of the rational paradigm, which is based on instrumental rationality and focuses on the effective use of resources, belong to the 'first' and 'second' generations of evaluation methods, as categorized by Guba and Lincoln. These methods include measurements and goal achievement models (Khakee et al. 2008). Communicative planning evaluation corresponds to what Guba and Lincoln (1989) describe as 'fourth generation evaluation'. In evaluation according to the communicative approach, focus is not only on effectiveness and legitimacy but also on democratic principles, integrity, mutual understanding, and consensus building (Khakee et al. 2008). Overall, however, planning and policy evaluation methods have a strong tendency toward quantitative modeling for measuring and forecasting the input and outcomes of planning alternatives and policies (Miller 2008).

The recent increase in environmental concerns and the rise of sustainability discourse have been concurrent with a growing interest in environmental policy evaluation among academics and policy makers, and in civil society as a whole. In recent years, various theoretical approaches and methods have been employed for policy evaluation in general, and in the environmental field in particular. Some of the main approaches used in the field of environment and planning include (Crabbě and Leroy 2008; Khakee et al. 2008): needs analysis (Reviere 1996); program theory evaluation (Stame 2004); environmental impact assessment; social impact assessment (Becker 2003); cost-benefit analysis; cost effectiveness analysis; integrated criteria analysis (Levent and Nijkamp 2005); multi criteria analysis; multi-objective decision-making (Alexander 2001); goal free evaluation; case study evaluation (Yin 2003); a multi-model system of sustainability indicators (Lombardi and Curwell 2005); a meta-analytic approach for analyzing and comparing sustainable development policies (Bizzaro and Nijkamp 1998); community impact analysis (Lichfield 2001); and environmental justice evaluation (Miller 2008).

Because of its complexities and uncertainties, *climate change* poses new and often dramatic challenges for urban planning and environmental policies and theories in general, and evaluation and assessment methods in particular. Existing methods are unable to meet these challenges, and this is particularly true in the arena of urban policy. For the most part, the existing evaluation methods propose and apply a package of assessment criteria, which are not deriving from theoretical foundations that are related to climate change (Mermet et al. 2010). Moreover, while evaluation of planning scenarios, policies and programs "has become a principle means of informing and supporting decision making" (Miller and Patassani 2005: xv). Moreover, while our urban phenomena have a 'qualitative nature' (Portugali 2010), however, environmental and policy evaluation methods have a strong tendency toward quantitative modeling for measuring and forecasting the input and outcomes of planning alternatives and policies (Miller 2008; Khakee et al. 2008).

Importantly, this paper argues that evaluation criteria that are not derived from a unified theory ultimately lack the theoretical and methodological consistency required to govern the evaluation method. Moreover, this paper recommends formulating theoretical foundations, or a conceptual framework for countering climate change *preceding to* establishing evaluation criteria or indicators, and on that basis to develop the concepts of evaluation. In other words, the formulation of a theoretical or conceptual framework must be understood as a necessary precondition for the construction of an evaluation method. This paper proposes a theory-based evaluation method for understanding policies and programs that aim to counter the deterioration of climate change and to coop with its uncertainties and risks.

4.3 The Countering Climate Change Evaluation Method (CCCEM)

As mentioned, the proposed evaluation method is a theory-based approach. Therefore, prior to building the evaluation method, a *conceptual framework for countering climate change* (CFCCC) was constructed. CFCCC is defined as a network or plane of interlinked concepts that together provide a comprehensive understanding of the phenomenon of fighting and coping with climate change and its results. The CFCCC consists of six concepts that together provide the theoretical foundations for effectively countering climate change phenomena. These concepts were identified through conceptual analysis of the literature on climate change. The previous chapter presents this conceptual framework. Together, these concepts— each of which represents a distinctive theme in the literature on climate change mitigation and adaptation—form the conceptual framework of the method. It is important to emphasize that the *conceptual framework* is not a mere collection of concepts. Moreover, CFCCC is a dynamic framework that may be revised according to new emerging studies and insights that renew and update our conventions (Fig. 4.1).

Consistently, in order to assess the role and contribution of each of the framework's concepts in countering climate change in the urban level, this chapter suggests the following procedures:

- Each concept has some components (sub-concepts). Every concept has components and there is no concept with only one component.
- Each component can be measured on a scale that ranges from vey low contribution in countering climate change to a very high contribution in countering climate change.
- A component may be measured both qualitatively and quantitatively, depending on its definition and the availability of data.
- Overall, a specific concept's contribution to countering climate change of a city is the sum of the contribution of its components.
- For the sake of uniformity, scales and measurements will be normalized and standardized.

The evaluation procedure involves applying each concept of assessment, along with its evaluative measures, to the plan under consideration. For example, when applying the concept of *equity*, we ask whether the plan addresses issues of environmental justice: whether it facilitates systematic public participation and whether it addresses the needs of different communities in the face of climate change.

The method is dynamic and evolutionary in character. It is based on collaboration with a large number of urban actors (including communities) and a variety of economic, social, environmental, and municipal entities, and is not based on the demanding and closed collaboration of client and evaluator. It is grounded in an inquiry paradigm and an interpretive or hermeneutic approach (see Guba and Lincoln 1989: 11–13). Because it is qualitative and employs no complicated

Fig. 4.1 The iterative process of building an assessment method

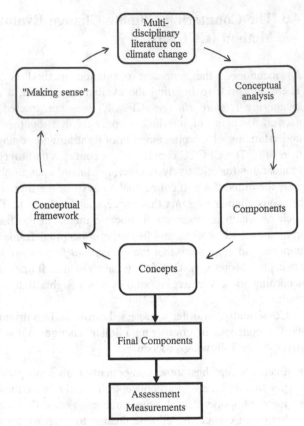

models, it is easy to grasp by practitioners, policy makers, and members of the public.

Formulation of this method was based on a number of shared values that underlie the conceptual framework for countering climate change (and, subsequently, the evaluation method as well). These values are extracted from the local and global discourse on countering climate change, and call for easing, reducing and eradicating, if possible, any human-made element that contributes to the deterioration of climate change, and for mitigating the impact of climate change on humans and other forms of life. These basic action-oriented values can serve as common ground for urban residents, communities, and other political, social, and economic stakeholders. If we share these basic values, or "shared constructions" in the words of Guba and Lincoln (1989: 8), we can 'make sense' of the evaluation outcomes. Moreover, "evaluation must have an action orientation that defines a course to be followed, stimulates involved stakeholders to follow it, and generates and preserves their commitment to do so" (Guba and Lincoln 1989: 8).

The countering climate change evaluation method is an integrated assessment approach that takes into account spatial, physical, economic, environmental, ecological, and social factors with the potential to help counter urban climate change. It provides a theoretical-based, coherent, and integrated method to aid in the structuring of present knowledge; to promote a broad view of the issues related to urban climate change; to help cities search more efficiently for possible responses to the uncertainties and to climate change's social, spatial, and economic effects on cities; to provide a consistent representation of the current uncertainties; and to address fundamental policy question regarding cities' responses and contributions to global climate change (Bruce et al. 1996).

In accordance with the work of Määttä and Rantala (2007), the role of the evaluator in the proposed method is threefold: the evaluator as *social engineer*, seeking out mechanisms that have objectively measurable outcomes; the evaluator as *program promoter*, serving as a counselling resource for community developers and other stakeholders; and the evaluator as *critical interpreter*, as reflected in the sociological intervention approach.

4.4 The Concepts of Assessment

The proposed *conceptual framework of evaluation* is composed of eight concepts of assessment, all of which are directed in one way or another toward coping with climate change and its results. The concepts are (Table 4.1):

4.4.1 Concept 1: Utopian Vision

Visions have significant function in the discourse and practice of planning that aims at achieving sustainability and coping with climate change impacts. It is important to mention that visionary frames are essential in climate change, as they serve to identify problematic conditions and the need for change, to propose future alternatives, and to urge all stakeholders to act in concert to affect change. Climate change planning visions should provide people with an interpretive framework that enables them to comprehend how the issue is related to their own lives in the present and future, and to the world at large (Taylor 2000; Benford and Snow 2000: 614). This concept assesses the visionary aspects of planning policies regarding the future of a city or a community, and if the frames of 'climate change' are central to the vision itself. What is the scope and nature of the vision; whether it related to climate change risk and uncertainties, and what are its dimensions; whether it includes social dimensions or environmental per se? This includes the future target of the reduction level of greenhouse emission in the city.

Table 4.1 The evaluation concepts for countering climate change

Concepts	Components	Key questions (measurements)
1: *Vision*		What is the scope and nature of the vision: whether it related to climate change risk and uncertainties, and what are its dimensions: whether it includes social dimensions or environmental per se?
2: Adaptation	C1. Uncertainties	C1. *Uncertainties*—whether the city and its plans and planning map draw the scenarios of uncertainties that may affect our cities as much as possible
	C2. Measures	C2. *Material measures*—what are the measures to quickly adapt to our rapidly changing environment and risk?
	C3. Urban vulnerability matrix	C3.1. *Demography*—what is the nature of the vulnerability demographic in the city by age, gender, health, and other social group?
		C3.2. *Spatiality*—what is the spatial distribution of risks, uncertainties, vulnerability and vulnerable communities in the city?
		C3.3. *Informality*—what is the scope, geography, socio-economic, demographic, and physical characters of existing informal settlements in or close to the city?
3: Equity—justice		C1: Who participates in decision-making and planning the *risk city*: coping with risk and threats, and planning for uncertainties?
4: Urban governance		C2: Is the urban governance approach integrating institutional, legal, social, economic, and environmental aspects?
5: Mitigation	C1 Energy	C1: How the city addresses the energy sector and whether it proposes strategies to reduce energy consumption and to use new alternative and cleaner energy sources?
	C2 Natural capital	C2.1 What are the existing and planned levels of GHG emissions?
		C2.2 What is the existing and planned state of the material resources, such as lands, green and agricultural areas close to the city, etc.?
	C3 *Renewable energy*	C3 What are the 'clean energy' or renewable energy targets and policies?
	C4 *Eco-form*	*C1 Compactness*: the level of compactness in the city
		C2: *Sustainable transport*: the existing and planned sustainable modes of transportation
		C3: *Density*: the existing and planned density and their nature of strategies
		C4: *Mixed land uses*: the diversity of land uses at city and community levels
		C5: *Diversity*: the characteristics of diversity at the social and housing level
		C6: *Passive solar design*: strategies and codes for green planning and design

(continued)

Table 4.1 (continued)

Concepts	Components	Key questions (measurements)
		C7: *Greening*: the nature and scope of greening the city
		C8: *Renewal and utilization*: the strategies and scope of cleaning and reasoning of contaminated areas
		C9: *Planning scale*: the planning level: a building, a block, a community, a neighbourhood, the city
6: Ecological economics		C1: What is the nature of the existing and planned ecological economy?

4.4.2 Concept 2: Equity

As mentioned in previous chapter, equity represents the social issues of climate change oriented planning policies. This concept is used to evaluate the social aspects of climate change related policies, including: environmental justice; public participation; and methods of addressing the vulnerability levels and scopes of a city and its individual community and neighbourhood.

Each urban society contains individuals and groups who are more vulnerable than others and lack the capacity to adapt to climate change (Schneider et al. 2007). The impacts of climate change are "socially differentiated," and are therefore a matter of urban distributional equity and justice, and climate change may lead to injustice and inequality, which may harm urban communities in general, and specific social groups. Injustice related to climate change may occurs along ethnic, gender, class, and racial lines, and even emerges among neighbourhoods and communities (see: Mohai et al. 2009; Adger 2001: 929; Bruce et al. 1996; Davies et al. 2008; IPCC 2007; Kasperson and Kasperson 2001; O'Brien et al. 2004; Paavola and Adger 2006; Tearfund 2008; Mohai et al. 2009; Adger et al. 2006). In addition, communities that are most vulnerable to climate change impact are usually those who live in more vulnerable, high-risk locations and may lack skills and adequate infrastructure and services (Satterthwaite 2008). Vulnerability refers to the "degree to which a system is susceptible to, and unable to cope with, adverse effects of climate change, including climate variability and extremes. Vulnerability is a function of a system's exposure, its sensitivity, and its adaptive capacity" (CCC 2010: 61). A society's development path, physical exposure, resource distribution, social networks, government institutions, and technological development influence its vulnerability (IPCC 2007).

4.4.3 Concept 3: Mitigation

This concept assesses the variety, scope, inclusiveness, and the nature of the measure that are undertaken to reduce GHG emissions. Mitigation refers to an "action to reduce the sources (or enhance the sinks) of factors causing climate change, such as greenhouse gases" (CCC 2010: 61), and to "the reduction of GHG emissions and their capture and storage in order to limit the extent of climate change" (Bulkeley 2010: 2.2). This concept is composed of two main components: *natural capital* and *urban eco-form* as follows:

4.4.3.1 Natural Capital

This component evaluates the consumption and—equally as important—the renewal of natural assets that are used for development, such as land, water, air, and open spaces. Significantly, it assesses the level of GHG emissions. A clean air is a highly significant component of countering climate change impacts. Natural capital refers to "the stock of all environmental and natural resource assets, from oil in the ground to the quality of soil and groundwater, from the stock of fish in the ocean to the capacity of the globe to recycle and absorb carbon" (Pearce et al. 1990: 1). Maintaining constant natural capital is an important criterion for sustainability (Pearce and Turner 1990: 44; Neumayer 2001; Geldrop and Withagen 2000). The stock of natural capital should not decrease, as this could endanger the ecological system and threaten the ability of future generations to generate wealth and maintain their well-being.

4.4.3.2 Ecological (Renewable) Energy

This component assesses what are the 'clean energy' or renewable energy targets (comparing to national and international recent requirements), and how the planning policies promote achieving the renewable energy objectives on the urban level. It evaluates how a plan addresses the energy sector and whether it proposes strategies to reduce energy consumption and to use new alternative and cleaner energy sources.

This component suggests that energy should be based on renewable energy and new low-carbon technologies in order to meet the target of emissions reduction. Energy is the most critical component of mitigation that aims to achieve sustainability and promote climate change oriented planning policies. Truly, *energy* is a "defining issue of our time" (Yumkella 2009, 1). The clean, renewable, and efficient use of energy is a central theme in all planning for the achievement of climate change objectives (UNIDO 2009: 6).

4.4.3.3 Eco-Form

This component evaluates spatial planning, architecture, design, and the ecologically desired form of the city and its components. The spatial and physical form of a city affects risk and issues of climate change in the city. It affects its ecosystems, the everyday activities and spatial practices of its inhabitants, and, eventually, climate change. Importantly, one of the unanswered questions of the adaptation-mitigation conflict is the question of what is the desirable density and compactness that are good for reducing risk and saving energies.

Jabareen (2006) suggests the following set of nine planning typologies, or components of evaluation, which are helpful in evaluating plans from the perspective of eco-form:

1. **Compactness**: it refers to urban intensity, contiguity, and connectivity of the urban form, and the way we plan and develop. Compact urban spaces minimize the use of energy, lands, and other resources. Intensification, a major strategy for achieving compactness, uses urban land more efficiently by increasing the density of development and activity.
2. **Sustainable Transport**: it suggests that planning should promote sustainable modes of transportation through traffic reduction; trip reduction; the encouragement of non-motorized travel (such as walking and cycling); transit-oriented development; safety; equitable access for all; and renewable energy sources, (Cervero 1998; Clercq and Bertolini 2003).
3. **Density**: High density is supposed to save significant amounts of energy and lands (Carl 2000; Walker and Rees 1997; Newman and Kenworthy 1989). Yet, density as well as compactness is problematic somehow to adaptation strategies.
4. **Mixed Land Uses**: this indicates the diversity of functional land uses, such as residential, commercial, industrial, institutional, and transportation. It allows planners to locate compatible land uses in close proximity to one another in order to decrease the travel distance between activities. This encourages walking and cycling and reduces the need for car travel, as jobs, shops, and leisure facilities.
5. **Diversity**: it is "a multidimensional phenomenon" that promotes other desirable urban features, including a larger variety of housing types, building densities, household sizes, ages, cultures, and incomes (Turner and Murray 2001: 320). Diversity is vital for cities. Without it, the urban system declines as a living place (Jacobs 1961).
6. **Passive Solar Design**: aims to reduce energy demands and to provide the best use of passive energy through specific planning and design measures, such as orientation, layout, landscaping, building design, urban materials, surface finish, vegetation, and bodies of water.
7. **Greening**: or bringing nature into the city, makes positive contributions to many aspects of the urban environment, including biodiversity; the lived-in urban environment; urban climate; economic attractiveness; community pride; and health and education.
8. **Renewal and Utilization**: Cleaning, rezoning, and developing contaminated sites are key aspects of revitalizing cities and neighbourhoods and contribute to their sustainability and to a healthier urban environment.
9. **Planning Scale**: influences and is influenced by climate change. For this reason, desirable planning scale should be considered and integrated into plans on the

regional, municipal, district, neighbourhood, street, site, and building levels. Planning that moves from the macro level to the micro level has a more holistic and positive impact on climate change.

4.4.4 Concept 4: Adaptation

This concept assesses planning adaptation strategies and policies and the planning strategies for addressing future uncertainties stemming from climate change. Does the plan include development projects for infrastructure design in order to reduce vulnerabilities and make the city more resilient? Does the plan enhance the city's adaptive planning capacity, or the ability of the planning system to respond successfully to climate variability and change? And is the plan flexible in order to adapt to rapidly changing environment?

4.4.4.1 Uncertainty

This component has a critical impact on urban vulnerability and requires the assessment of environmental risks and hazards that are difficult to predict but must be taken into account in city planning and risk management. Uncertainty is about a lack of knowledge (Abbot 2009: 503). Apparently, vulnerability assessments often ignore the non-climatic drivers of future risk (Storch and Downes 2011) and uncertainties. Therefore, this critical component assesses whether the city and its plans and planning map and draw the scenarios of uncertainties that may affect our cities as much as possible.

4.4.4.2 Material Measures

This component is literally about material measures that are related to adaptation. What are these measures, and are they flexible enough to quickly adapt to our rapidly changing environment? Planning must develop a better understanding of the risks that climate change poses to infrastructure, households, and communities. To address these risks, planners have two types of uncertainties or adaptation managements at their disposal: (1) Ex-ante management, or actions taken to reduce and/ or prevent risky events; and (2) Ex-post management, or actions taken to recover losses after a risky event (Heltberg et al. 2009).

4.4.4.3 Urban Vulnerability Matrix

This component addresses how hazards, risks, and uncertainties affect various urban communities and urban groups. Therefore, the role of the Vulnerability Analysis

Matrix is to analyze and identify the type, demography, intensity, scope, and spatial distribution of environmental risk, natural disasters, and future uncertainties in cities.

The Vulnerability Analysis Matrix is composed of three main components that determine its scope, environmental, social, and spatial nature. These four components are:

Demography of Vulnerability This component assesses and examines the demographic and socio-economic aspects of urban vulnerability. It assumes that there are individuals and groups within all societies who are more vulnerable than others and lack the capacity to adapt to climate change (Schneider et al. 2007: 719). Demographic, health, and socio-economic variables affect the ability of individuals and urban communities to face and cope with environmental risk and future uncertainties. These variables affect the mitigation of risk, response and recovery from natural disasters.

Informality This component assesses the scale and social, economic, and environmental conditions of informal urban spaces and their vulnerability. Informal spaces are unplanned, chaotic, and disorderly and it is assumed that the scale and human condition of informal places within a city have a significant impact on its vulnerability.

Spatial Distribution of Vulnerability This component assesses the spatial distribution of risks, uncertainties, vulnerability and vulnerable communities in cities. Environmental risks and hazards are not always evenly distributed geographically, and some communities may be affected more than others. Mapping the spatial distribution of risks and hazards is critical for planning and management at the present and for the future.

4.4.5 Concept 5: Integrative Approach

This concept evaluates the integrative framework for city planning and adaptive management under conditions of uncertainty, and the spectrum of collaboration that a plan proposes. It is assumed that in order to enhance the urban governance of the *risk city* we need to expand and improve the local capacity through increasing knowledge, providing resources, establishing new institutions, enhancing good governance, and granting more local autonomy.

4.4.6 Concept 6: Ecological Economics

This concept evaluates the planning economic aspects that aim to achieve clean energy targets and the incentives and economic engines it puts in place to meet climate change objectives. The major assumption behind this concept is that environmentally sound economics can play a decisive role in achieving climate change objectives in a capitalist world. Therefore, urban planning and public policies should stimulate markets for 'green' products and services, promote environmentally friendly consumption, and contribute to urban economic development by creating a cleaner environment (Hsu 2006: 11).

4.5 Conclusions

This chapter represents an interdisciplinary evaluation method for evaluating urban climate change policies and assessing their potential contribution to increased trust and decrease risk in the *risk city*, as well to promote sustainability, more effective responsive and adaptive measures, and provisions to contend with climate change and its implications for the built environment.

This chapter elaborated the *Countering Climate Change Evaluation Method* (CCCEM). CCCEM has some important characteristics as follows:

1. CCCEM acknowledges and thereafter addresses the multidisciplinary and complexity nature of countering climate changes in the urban level. Cities and their socio-spatial and environmental phenomena are complex and therefore they need multidisciplinary methods to capture their complexity. Therefore, CCCEM is a complexity evaluation method, and its concepts are interlinked and have mutual affects. In other words, each concept is affected and affects the remainder of the concepts. There are contradictions between practices, such as between the desired eco-form for mitigation and the desired form for adaptation, which for example seeks in many cases less dense urban paces.

2. CCCEM proposes easy to grasp methods. Climate change is a complex broad scale public issue that the assessment method proposed in this paper helps 'make sense' of for civil society, the private sector, practitioners, policy makers, and the public at large. The proposed conceptual framework and assessment method, when applied as a whole, provides an informative, easy to grasp, effective, and constructive means of evaluating urban plans and illuminating their strengths and weaknesses. In accordance with recent critique on the limited application of complexity theories in the urban context, the proposed study acknowledges the essentially '*qualitative nature*' of urban phenomena and makes an innovative contribution to the scholarship by applying an innovative, multidisciplinary *qualitative* methodology to complexity theories in countering climate change in the urban context.

3. CCCEM proposes a dynamic and flexible conceptual framework and methodology that addresses uncertainties and is able to accommodate new emerging risks that may affect cities in a dramatic way, integrating them into countering climate change framework.

4. CCCEM proposes systematic manners that allow us to compare also between cities in terms of countering climate change.

5. Due to its 'easy to grasp' and 'making sense' qualities, CCCEM has the potential to facilitate more *awareness* among scholars, professionals, decision makers, and the public as a whole regarding the current and future direction of cities regarding climate change issues. The proposed multifaceted conceptual framework will help us determine what needs to be done to increase the resilience of our cities, thereby enabling us to work more effectively toward making them more safe and secure.

References

Abbott, J. (2009). Planning for complex metropolitan regions: A better future or a more certain one? *Journal of Planning Education and Research, 28*, 503–517.

Adger, W. N. (2001). Scales of governance and environmental justice for adaptation and mitigation of climate change. *Journal of International Development, 13*(7), 921–931.

Adger, W. N., Paavola, J., Huq, S., & Mace, M. J. (2006). *Fairness in adaptation to climate change*. Cambridge, MA: MIT Press.

Alexander, E. (2001). Unvaluing evaluation: Sensitivity analysis in MODM application. In H. Voogd (Ed.), *Recent development in evaluation* (pp. 319–340). Groningen: Geo Press.

Baycan T., & Nijkamp, P. (2005). Evaluation of urban green spaces. In D. Miller & D. Patassini (Eds.), *Accounting for non-market values in planning evaluation* (pp. 63–88). Farnham: Ashgate Publication.

Becker, H. A. (2003). *The international handbook of social impact assessment: Conceptual and methodological advances*. Cheltenham, UK: Edward Elgar Publishing.

Benford, R. D., & Snow, D. A. (2000). Framing Processes and Social Movements: An Overview and Assessment. *Annual Review of Sociology, 26*, 611–639.

Bizzaro, F., & Nijkamp, P. (1998). Cultural heritage and the urban revitalization: A meta-analytic approach to urban sustainability. In N. Lichfield, A. Barbanenete, D. Borri, A. Khakee, & A. Prat (Eds.), *Evaluation in planning facing the challenges of complexity* (pp. 193–212). Dordrecht: Kluwer.

Bruce, J. P., Lee, H., & Haites, E. F. (Eds.). (1996). *Climate change 1995: Economic and social dimensions of climate change: Contribution of working group III to the second assessment report of the intergovernmental panel on climate change*. Cambridge: Cambridge University Press.

Bulkeley, H., & Newell, P. (2010). *Governing climate change*. London: Routledge.

Carl, P. (2000). Urban density and block metabolism. In K. Steemers & S. Yannas (Eds.), *Proceedings of PLEA 2000 Architecture, City, Environment* (pp. 343–347). London: James & James.

CCC—Committee on Climate Change Adaptation. (2010). *How well prepared is the UK for climate change?* www.theccc.org.uk.

Cervero, R. (1998). *The transit metropolis: A global inquiry*. Washingdon, D.C.: Island Press.

Clercq, F., & Bertolini, L. (2003). Achieving sustainable accessibility: An evaluation of policy measures in the Amsterdam area. *Built Environment, 29*(1), 36–47.

Crabbě, A., & Leroy, P. (2008). *The handbook of environmental policy evaluation*. London: Earthscan.

Davies, M., Guenther, B., Leavy, J., Mitchell, T., & Tanner, T. (2008). *Climate change adaptation, disaster risk reduction and social protection: Complementary roles in agriculture and rural growth?* Institute of Development Studies Centre for Social Protection and Climate Change and Disasters Group. IDS: Institute of Developing Studies. http://www.climategovernance.org/docs/SP-CC-DRR_idsDFID_08final.pdf.

Diez, M. A. (2001). The evaluation of regional innovation and cluster policies: Towards a partipatory approach. *European Planning Studies, 9*(7), 907–923.

Geldrop, J., & Withagen, C. (2000). Natural capital and sustainability. *Ecological Economics, 32* (3), 445–455.

Guba G. E., & Lincoln S. Y. (1989). *Forth generation evaluation.* California: Sage Publications.

Hallegatte, S., Przyluski, V., & Vogt-Schilb, A. (2011). Building world narratives for climate change impact, adaptation and vulnerability analyses. *Nature Climate Change, 1,* 151–155.

Heltberg, R., Siegel, P. B., & Jorgensen, S. L. (2009). Addressing human vulnerability to climate change: Toward a 'no-regrets' approach. *Global Environmental Change, 19*(2009), 89–99.

Hsu, D. (2006). *Sustainable New York city. New York city: Design trust for public space and the New York City office of environmental coordination.* New York City: New York City Office of Environmental Coordination.

IPCC—Intergovernmental Panel on Climate Change. (2007). *Climate change 2007: Fourth assessment report of the intergovernmental panel on climate change.* Cambridge, MA: Cambridge University Press.

Jabareen, Y. (2006). Sustainable urban forms: Their typologies, models, and concepts. *Journal of Planning Education and Research, 26*(1), 38–52.

Jacobs, J. (1961). *The death and life of great American cities.* New York: Random House.

Kasperson, R. E., & Kasperson, J. X. (2001). *Climate change, vulnerability and social justice.* Stockholm: Stockholm Environment Institute.

Khakee, A., Hull, A., Miller, D., & Woltjer, J. (2008). Introduction. In A. Khakee, A. Hull, D. Miller & J. Woltjer (Eds.), *New principles in planning evaluation* (pp. 1–16). Farnham: Ashgate.

Levent, T. B., & Nijkamp, P. (2005). Evaluation of urban green spaces. In D. Miller & D. Patassani (Eds.), *Beyond benefit cost analysis. Accounting for non-market values in planning evaluation* (pp. 63–88). Aldershot: Ashgate.

Lichfield, N. (2001). The philosophy and role of community impact evaluation in the planning system. In H. Voogd (Ed.), *Recent development in evaluation* (pp. 153–174). Groningen: Geo Press.

Lombardi, P., & Curwell, S. (2005). Analysis of the INTELCITY Scenarios for the city of future from a southern European perspective. In D. Miller & D. Patassani (Eds.), *Beyond benefit cost analysis. Accounting for non-market values in planning evaluation* (pp. 207–224). Aldershot: Ashgate.

Määttä, M., & Rantala, K. (2007). The evaluator as a critical interpreter: Comparing evaluations of multi-actor drug prevention policy. *Evaluation, 13*(4), 457–476.

Mermet, L., Billé, R., & Leroy, M. (2010). Concern-focused evaluation for ambiguous and conflicting policies: An approach from the environmental field. *American Journal of Evaluation, 31,* 180–198.

Miller, D. (2008). Methods for assessing environmental justice in planning evaluation-an approach and an application. In A. Khakee, A. Hull, D. Miller & J. Woltjer (Eds.), *New principles in planning evaluation* (pp. 19–33). Farnham: Ashgate.

Miller, D., & Patassini, D. (Eds.) (2005). *Accounting for non-market values in planning evaluation* (pp. 63–88). Farnham: Ashgate Publication.

Mohai, P., Pellow, D., & Roberts, J. T. (2009). Environmental justice. *Annual Review of Environment and Resources, 34,* 405–430.

Neumayer, E. (2001). Do countries fail to raise environmental standards? An evaluation of policy options addressing regulatory chill. *International Journal of Sustainable Development, Inderscience Enterprises Ltd, 4*(3), 231–244.

Newman, P., & Kenworthy, J. (1989). Gasoline consumption and cities: a comparison of US cities with a global survey. *Journal of the American Planning Association, 55,* 23–37.

O'Brien, K., Leichenko, R., Kelkar, U., Venema, H., Aandahl, G., Tompkins, H., et al. (2004). Mapping vulnerability to multiple stressors: Climate change and globalization in India. *Global Environmental Change, 14,* 303–313.

Paavola, J., & Adger, W. N. (2006). Fair adaptation to climate change. *Ecological Economics, 56* (4), 594–609.

Pearce, D., & Turner, R. K. (1990). *Economics of natural resources and the environment.* Baltimore: Johns Hopkins University Press.

Pearce, D., Barbier, E., & Markandya, A. (1990). *Sustainable development: Economics and environment in the third world.* London: Earthscan Publications.

Portugali, J. (2010). Complexity, cognition and the city. Berlin: Springer (in press).

Reviere, R. (Ed.). (1996). *Needs assessment: A creative and practical guide for social scientists.* Washington, D.C.: Taylor & Francis.

Satterthwaite, D. (2008). *Climate change and urbanization: effects and implications for urban governance.* Presented at UN Expert Group Meeting on Population Distribution, Urbanization, Internal Migration and Development. UN/POP/EGMURB/2008/16/.

Schneider, S. H., Semenov, S., Patwardhan, A., Burton, I., Magadza, C. H. D., Oppenheimer, M., et al. (2007). Assessing key vulnerabilities and the risk from climate change. Climate change 2007: Impacts, adaptation and vulnerability. In M. L. Parry, O. F. Canziani, J. P. Palutikof, P. J. van der Linden, & C. E. Hanson (Eds.), *Contribution of working group II to the fourth assessment report of the intergovernmental panel on climate change* (pp. 779–810). Cambridge, UK: Cambridge University Press.

Stame, N. (2004). Theory-based evaluation and types of complexity. *Evaluation, 10*(1), 58–76.

Storch, H., & Downes, N. K. (2011). A scenario-based approach to assess Ho Chi Minh City's urban development strategies against the impact of climate change. *Cities, 28*(6), 517–526.

Taylor, D. E. (2000). The rise of the environmental justice paradigm: Injustice framing and the social construction of environmental discourses. *American Behavioral Scientist, 43,* 508–580.

Tearfund. (2008). *Linking climate change adaptation and disaster risk reduction.* Web White Pap. http://www.tearfund.org/webdocs/Website/Campaigning/CCAandDRRweb.pdf.

Turner, S. R. S., & Murray, M. S. (2001). Managing growth in a climate of urban diversity: South Florida's Eastward ho! Initiative. *Journal of Planning Education and Research, 20,* 308–328.

UNIDO—United Nations Industrial Development Organization. (2009). *Energy and climate change: Greening the Industrial agenda.* http://www.unido.org.

Vedung, E. (2010). Four waves of evaluation diffusion. *Evaluation, 16*(3), 263–277.

Walker, L., & Rees, W. (1997). Urban density and ecological footprints—an analysis of Canadian households. In M. Roseland (Ed.), *Eco-city dimensions: Healthy communities, healthy planet.* Canada: New Society Publishers.

Yin, R. (2003). *Case study research: Design and methods.* Thousand Oaks, CA: Sage.

Yumkella, K. K. (2009). *Forward. in energy and climate change: Greening the industrial agenda* (P. 1). UNIDO-United Nations Industrial Development Organization, 2010. http://www.unido. org.

Chapter 5
Contemporary Planning of the Risk City: The Case of New York City

5.1 Introduction

Actually, cities and city planning have an important role to play in contending with the future impacts of climate change, which, with its many complexities and uncertainties, poses new challenges for the planning profession. For this reason, generating a broad understanding of planning's role in fighting climate change is now emerging as a crucial task for planners, who are still in the preliminary stages of setting agendas and exploring possible directions (Kern and Alber 2008; Priemus and Rietveld 2009; van Leeuwen et al. 2009; Swart et al. 2009). Swart et al. (2009: 152) argue that although spatial planning is considered to be an essential lever of adaptation policy, "the references made to planning instruments widely remains very generic and vague" and "the specific potential of planning tools is, therefore, not yet addressed." Moreover, even though planners and planning have the "ability to think the big, long-term thoughts about the interrelatedness and interdependency" of complex issues such as hazard mitigation and climate change (Schwab 2010: 5), many cities around the world, including the most pioneering among them, still fail to utilize comprehensive and spatial planning in their fight against climate change (Kern and Alber 2008).

Many cities and communities, especially in the West, are now grappling with climate change through a multitude of practices aimed at mitigating greenhouse emissions and adapting to the anticipated, albeit uncertain, impacts of climate change (Jabareen 2006, 2008). Recently, we have become increasingly aware of environmental degradation and the huge risks and uncertainties that climate change poses to our cities and communities. Climate change is likely to affect the social, economic, ecological, and physical systems of every city. Ironically, cities themselves, through their economic production and modes of consumption, are major contributors to the same environmental crisis from which they will suffer.

© Springer Science+Business Media Dordrecht 2015
Y. Jabareen, *The Risk City*, Lecture Notes in Energy 29,
DOI 10.1007/978-94-017-9768-9_5

In respect to climate change and sustainability, NYC "set forth an ambitious effort" to push forward many efforts under the umbrella of the *PlaNYC 2030* (Solecki 2012: 570). Some scholars suggest that the *PlaNYC* is an ambitious and landmark sustainability plan aiming at charting the city's future for the coming decades, and addressing the challenges of climate change (Rosenzweig and Solecki 2010b: 19; Rosan 2012).

PlaNYC "is the sustainability and resiliency blueprint for New York City" (PlaNYC 2014). It is a long-term plan for the city (Angotti 2010). A fundamental assumption of *PlaNYC*, which was launched on Earth Day 2007, is that "climate change poses real and significant risks to New York City" (PlaNYC: Progress Report 2009: 39). Rosenzweig et al. (2010) suggest "New York has won considerable recognition for its long-term growth and sustainability plan, *PlaNYC 2030*". *PlaNYC*, which was first implemented in 2007 and updated in 2011, set an ambitious goal of a 30 % reduction of GHG emissions from 2005 levels by 2030 (Solecki 2014).

PlaNYC is an inclusive plan for a big city. The present population of New York City is approximately 8,363,700 people (US Census Bureau 2009), and according to the plan, the target population for 2030 will surge past nine million (PlaNYC: 6). The work force will grow by 750,000 jobs and the need for 60 million square feet of additional commercial space, which the Plan suggests should be filled by the "re-emergence of Lower Manhattan and new central business districts in Hudson Yards, Long Island City and Downtown Brooklyn" (PlaNYC: 6). "The plan predicts 65 million visitors to the city by 2030. The additional jobs, tourists, and residents could generate an additional $13 billion annually—money that can be used to help fund some of the initiatives described in the following pages and to provide the services that our residents, businesses, workers, and visitors deserve" (PlaNYC: 6). The plan is composed of 127 new initiatives that aim to strengthen the economy, public health, and the quality of life in the city. Collectively, they will form the broadest attack on climate change ever undertaken by an American city. In addition, "most of *PlaNYC's* 127 separate initiatives contribute directly to achieving the city's GHG reduction goals: to reduce citywide GHG emissions by 30 % by 2030 and to reduce City government GHG emissions by 30 % by 2017" (PlaNYC: Inventory of New York City Greenhouse Gas Emission 2009).

The evaluation framework consists of six concepts of assessment that were identified through conceptual analyses of the planning and interdisciplinary literature on sustainability and climate change. Together, these concepts—each of which represents a distinctive field of practices—form the conceptual framework of the method. Each concept is composed of several criteria of evaluation.

5.2 The Assessment of PlaNYC

This section presents the assessment findings and insights of *PlaNYC* 2030, which are presented through six concepts of Countering Climate Change Evaluation Method (CCCEM). In addition, this section takes advantages of many reports that were prepared by New York City and New York State, such as: Progress Reports of *PlaNYC* 2030; Climate Change Reports; Energy Conservation Plans; Greenhouse Gas Inventory; Municipal Energy Conservation; Think Locally, Act Globally: How Curbing Global Warming can Improve Local Public Health; PlaNYC: Inventory of New York City Greenhouse Gas Emission; and NPCC—New York City Panel on Climate Change: Climate Risk Information.

5.2.1   The Utopian Vision of PlaNYC: The Problem Statement

Risk is at the departure point of the vision and problem statement of *PlaNYC*. Climate change played a major role in formulating the problems facing New York City and justifying the urgency of the new plan. Both *PlaNYC* and the NYC Panel on Climate Change (NPCC 2009), a public body proposed by *PlaNYC* and convened by the mayor of New York in 2008, portrays New York as a city at risk. Beyond doubt, "climate change has the potential to impact everyday life in New York City," which supposes to expose the city and its residents to new hazards and heightened risks (Rosenzweig and Solecki 2010b: 13). Moreover, The NPCC holds that "climate change poses a range of hazards to New York City and its infrastructure" and that "these changes suggest a need for the City to rethink the way it operates and adapts to its evolving environment" (NPCC 2009: 3).

From the outset, *PlaNYC diagnoses* the local and global climate change crisis as problematic and critical for New York City and the world as a whole. The Plan states that New York has "already started to experience warmer, more unpredictable weather and rising sea levels" and notes scientists' projections that, as temperatures rise across the globe toward the end of the century, New York City could find itself with between 40 and 89 days that are 90° or hotter each year. Climate change is likely to bring warmer temperatures to New York City and the surrounding region, as the mean annual temperatures projected by global climate models are expected to increase by 1.5°–3° (Fahrenheit) by the 2020s, 3°–5° by the 2050s, and 4°–7.5° by the 2080s (NPCC 2009). The city will also experience more intense rainstorms, while annual precipitation is likely to increase and droughts are likely to become more severe toward the end of twenty first century. Heat waves are also expected to become more frequent, intense, and longer in duration, and sea levels are likely to rise, with an increase of 2–5 in. by the 2020s, 7–12 in. by the 2050s, and 12–23 in. by the 2080s. "As a coastal city," PlaNYC concludes, "we are vulnerable to the most dramatic effects of global warming: rising sea levels and intensifying storms" (PlaNYC: 133). In contrast to the period preceding the industrial revolution when sea levels rose at rates of 0.34–0.43 in. per decade, current rates of increase around New York City range from 0.86 to 1.5 in. per decade (NPCC 2009: 5–9; Gehrels et al. 2005; Holgate and Woodworth 2004). As a result, general flooding and storm-related coastal flooding are likely to increase as well (NPCC 2009: 4). New York City has almost 578 miles of coastline and over half a million residents living within the current flood plain, and this poses a particularly dangerous risk to the city. NPCC holds that New York City already faces the probability of a "hundred year flood" once every 80 years. This could increase to once every 43 years by the 2020s and to once every 19 years by the 2050s. According to one estimate, a Category 2 hurricane would inflict more damage on New York than on any other American city except Miami (NPCC 2009, 8).

As a result, climate change poses particular threats to New York's infrastructure, including: increased summertime strain on materials; higher peak electricity loads in summer and reductions in heating capacity in winter; voltage fluctuations, equipment damage and service interruptions; increased demands on HVAC systems; transportation service disruption; increased street, basement and sewer flooding; reduction of water quality; inundation of low-lying areas and wetlands; increased structural damage and impaired operations; and increased need for emergency management procedures (NPCC 2009: 4–30).

The City's infrastructure adds dramatically to the uncertainties surrounding climate change. According to *PlaNYC* (2007: 7), the City's infrastructure "is the oldest in America." Not only are the subway system and highway networks heavily-used, but about 3000 miles of roads, bridges, and tunnels are in need of repair, as are many subway stations. To make matters worse, the water infrastructure has not been inspected in more than 70 years, and 52 % of the city's tributaries that run adjacent to the shoreline and pass through neighborhoods are unsafe even for boating. Finally, about 7600 acres throughout the boroughs remain contaminated, and the city suffers from one of the worst asthma rates in the country (PlaNYC, 7). Between the years 2000 and 2005, New York's greenhouse gas emissions increased by almost 5 % (PlaNYC, 135), which is of particular significance because New York City emits nearly 0.25 % of the world's total greenhouse gases. As a coastal city, PlaNYC concludes, "we are vulnerable to the most dramatic effects of global warming: rising sea levels and intensifying storms" (PlaNYC, 133).

Not only the City and its *PlaNYC* but also the State of New York presents its main problems based on risk of climate change and its uncertainties. The State Commission 2100 suggests that:

> There are significant climate change risks including sea level rise, changing patterns of precipitation, temperature change and increasingly frequent extreme weather events. There are demographic pressures, with significant population growth predicted for New York state, and structural changes within the population, including further urbanization, the growth of suburban poverty, as well as the continuing needs of those living below the poverty level and a growing aging population (NYS 2100 2013: 10).

In this way, climate change and its risk and effects play a decisive role in the manner in which *PlaNYC* describes the problems currently facing the city, which may lead to a dramatic deterioration of life in the city in the future.

Later, PlaNYC portrays New York City as the most sustainable and "one of the most environmentally efficient cities" in the USA (PlaNYC: 135), producing "less than a third of the CO_2e generated by the average American." In this way, it holds, "Growing New York is, itself, a climate change strategy." According to the Plan,

New York City is a globally responsible, pioneering, modern and innovative city—
a city with an "unending sense of possibility" (PlaNYC: 130). Still, PlaNYC
acknowledges, "in spite of our inherent efficiency, we can do better. And we must.
Instead we are doing worse" (PlaNYC: 135). As one of the world's most specta-
cular cities, planners hold, New York should seize the opportunity and "define the
role of cities in the 21st century and lead the fight against global warming"
(PlaNYC: 130). The City "cannot afford to wait until others take the lead" on
slowing climate change. "New York has always pioneered answers to some of the
most pressing problems of the modern age," the planners argue, and "it is
incumbent on us to do so again, and rise to the definitive challenge of the 21st
century" (PlaNYC: 9).

Climate change is central to the visioning of New York City according to *PlaNYC*.
PlaNYC's vision generates a sense of local and global *urgency*: "unless the public...
appreciate[s] the urgency...we will not meet our goal" (PlaNYC: 110). "Meanwhile,
we will face an increasingly precarious environment and the growing danger of
climate change that imperils not just our city, but the planet. We have offered a
different vision... It is a vision of New York as the first sustainable twenty-first
century city—but it is more than that. It is a plan to get there" (PlaNYC: 141).

The planning vision promises a *better future*: "The result, we believe, is the most
sweeping plan to strengthen New York's urban environment in the city's modern
history... we have developed a plan that can become a model for cities in the
twenty-first century" (PlaNYC: 10).

> It is a vision of providing New Yorkers with the cleanest air of any big city in the nation; of
> maintaining the purity of our drinking water;...; of producing more energy more cleanly
> and more reliably, and offering more choices on how to travel quickly and efficiently across
> our city. It is a vision where contaminated land is reclaimed and restored to communities;
> where every family lives near a park or playground; where housing is sustainable and
> available to New Yorkers from every background, reflecting the diversity that has defined
> our city for centuries (PlaNYC: 141).

PlaNYC casts "climate change," or "sustainability," as a major concern and central
theme of the plan. New York City Mayor Michael Bloomberg describes *PlaNYC* as
"a long-term vision for a sustainable New York City" which "has been acknowledged
around the world as one of the most ambitious—and most pragmatic—sustainability
plans anywhere" (*PlaNYC*: Progress Report 2009: 4). He also maintains that each of
the plan's 127 initiatives "will not only strengthen our economic foundation and
improve our quality of life; collectively, they will also form a frontal assault on the
biggest challenge of all: global climate change" (*PlaNYC*: Progress Report 2009: 2).

The vision advanced in *PlaNYC* includes solutions and planning strategies, calls
for collective action, and promises that "we can do better. Together, we can create a
greener, greater New York" (PlaNYC: 3). In the words of the mayor, "Truly,
PlaNYC has become a citywide effort...we are creating a better and more sustainable city—one that will rise above the current economic turmoil and show the

world how it is possible to come back stronger than ever... The City is committed to these goals, and together, I know we can build a greener, greater New York" (*PlaNYC*: Progress Report 2009: 4).

The vision of PlaNYC is ambitious: its practical aim is to reduce emissions by 30 %, and its physical agenda is to develop New York City as a "greener, greater New York." The vision adequately addresses local and global climate change as a central concern of planning and future development. It aims to inspire and mobilize New Yorkers to collectively adhere to the planning initiatives and to build consensus and legitimacy for its implementation. For this reason, the word "*we*" appears 1708 times in the 156 pages of *PlaNYC*, or about 11 times per page. Yet, the vision overlooks the social and cultural agenda of such a diverse city. Strikingly, even though New York is "more diverse than ever; today nearly 60 % of New Yorkers are either foreign-born or the children of immigrants" (PlaNYC: 4), with 174 languages spoken by the city residents, the vision neglects the social and cultural issues related to this majority of the city's population.

PlaNYC classifies its main goals under six main themes, as shown in Table 5.1. All of the plan's main themes, and nine of its ten goals, are physically and environmentally oriented, while only one goal, which focuses on housing and affordability, can be considered social in nature. In this way, climate change clearly plays a key role in the vision of the City and maintains a strong implicit and explicit presence within the plan's goals.

Table 5.1 Goals of *PlaNYC*

Main theme	Sub-theme	Goals
Land	Housing	*"Create homes for almost a million more New Yorkers, while making housing more affordable and sustainable"*
	Open space	*"Ensure that all New Yorkers live within a 10-minutes walk of a park"*
	Brownfields	*"Clean up all contaminated land in NYC"*
Water	Water quality	*"Open 90 % of our waterways to recreation by preserving natural areas and reducing pollution"*
	Water network	*"Develop critical backup systems for our aging water network to ensure long-term reliability"*
Transportation	Congestion	*"Improve travel times by adding transit capacity for millions more residents, visitors, and workers"*
	State of good repair	*"Reach a full 'state of good repair' on NYC's roads, subways, and rails for the first time in history"*
Energy	Energy	*"Provide cleaner, more reliable power for every New Yorker by upgrading our energy infrastructure"*
Air	Air quality	*"Achieve the cleanest air quality of any big U.S. city"*
Climate change	Climate change	*"Reduce global warming emission by 30 %"*

Source Based on information contained in *PlaNYC* pages 15, 51, 73, 99, 117, 131

5.2.2 *Adaptation in PlaNYC*

The adaptation concept contains three components: Uncertainties, *material measures,* and Urban Vulnerability Matrix.

5.2.2.1 Uncertainties

A fundamental assumption of PlaNYC, which was launched on Earth Day 2007, is that "climate change poses real and significant risks to New York City." (*PlaNYC*: Progress Report 2009: 39). New York city is portrayed as a city at risk by both PlaNYC and the New York City Panel on Climate Change (NPCC 2009), a public body proposed by PlaNYC (PlaNYC: 139) and convened by the mayor of New York in 2008 in order to achieve the climate change related goals outlined in the Plan. The NPCC holds that "climate change poses a range of hazards to New York City and its infrastructure" and that "these changes suggest a need for the City to rethink the way it operates and adapts to its evolving environment" (NPCC 2009: 3). According to the NPCC (2009), climate change is likely to bring warmer temperatures to New York City and the surrounding region, as mean annual temperatures, as projected by global climate models, increase by 1.5–3 °F by the 2020s, 3–5 °F by the 2050s, and 4–7.5 °F by the 2080s (NPCC 2009). In addition, the city will also see more intense rainstorms, while annual precipitation is likely to increase and droughts become more severe toward the end of 21st century. Heat waves are also likely to become more frequent, intense, and longer in duration. Furthermore, sea levels are also likely to rise in the decades to come, with rises of 2–5 in. by the 2020s, 7–12 in. by the 2050s, and 12–23 in. by the 2080s. In comparison to the period preceding the industrial revolution, when sea levels rose at rates of 0.34–0.43 in. per decade, current rates around New York City range between 0.86 and 1.5 in. per decade (NPCC 2009: 5–9; Gehrels et al. 2005; Holgate and Woodworth 2004). As a result, flooding and storm-related coastal flooding are likely to increase as well (NPCC 2009: 4). New York City has almost 578 miles of coastline and over half a million residents living within the current flood plain, which is especially dangerous to New York. In fact, at the current sea level, NPCC suggests that New York City already faces the probability of a "hundred year flood" once every 80 years. This could increase to once every 43 years by the 2020s and to once every 19 years by the 2050s. According to one estimate, a Category 2 hurricane would inflict more damage on New York than on any other American city except Miami (NPCC 2009: 8).

Climate change poses particular threats to the city's infrastructure, in the form of: increased summertime strain on materials; increased peak electricity loads in summer and reduced heating in winter; voltage fluctuations, equipment damage and service interruptions; increased demands on HVAC systems; transportation service disruption; increased street, basement and sewer flooding; reduction of water quality; inundation of low-lying areas and wetlands; increased structural damage and impaired operations; and increased need for emergency management procedures (NPCC 2009: 4–30).

In addition to these threats, the already deteriorating physical condition of city infrastructure adds dramatically to the uncertainties surrounding climate change. According to *PlaNYC* (2007: 7), New York City's infrastructure "is the oldest in America." Not only are the subway system and highway networks heavily-used, but about 3000 miles of roads, bridges, and tunnels are in need of repair, as are many subway stations. To make matters worse, the water infrastructure has not been inspected in more than 70 years, and 52 % of the city's tributaries that run adjacent to the shoreline and pass through neighborhoods are unsafe even for boating. Finally, about 7600 acres throughout the boroughs remain contaminated, and the city suffers from one of the worst asthma rates in the country (PlaNYC: 7).

With regard to these risks and uncertainties, *PlaNYC* explains that "there is no silver bullet to deal with climate change," and "as a result, our strategy to help stem climate change is the sum of all the initiatives in this plan" (PlaNYC: 135). The Plan's main thrust for climate change adaptation appears to lie in the creation of "an intergovernmental Task Force to protect our city's vital infrastructure" and "to work with vulnerable neighborhoods to develop site-specific strategies" (PlaNYC: 136). In addition, *PlaNYC* proposes the establishment of a New York City Climate Change Advisory Board, a citywide strategic planning process "to determine the impacts of climate change to public health and other elements of the City and begin identifying viable adaptation strategies" (*PlaNYC*: Progress Report 2009: 39). Proposed adaptation policies also include measures to fortify the city's critical infrastructure, to be implemented through close cooperation between city, state, and federal agencies and authorities; updating the flood plain maps to better protect areas that are most vulnerable to flooding; and working with at-risk neighborhoods across the city to develop site specific plans. "In addition to these targeted initiatives," the Plan reads, "we must also embrace a broader perspective, tracking the emerging data on climate change and its potential impacts on our city" (PlaNYC: 136).

PlaNYC addresses future uncertainties of climate change on a general level and suggests primarily institutional procedures—establishment of the NPCC and the Climate Change Advisory Board—to meet the challenge. Without a doubt, these bodies, which are charged with monitoring climate change parameters vis-à-vis the City and proposing adjustment policies, enhance the city's urban adaptive planning capacity. At the same time, however, the plan's adaptation strategy is based principally on emission reduction, an *ex-ante* strategy. In this way, *PlaNYC* fails to prepare the city and its infrastructure for the disasters that could stem from climate change. For example, the Plan proposes no infrastructure design or development projects along the city's vulnerable 570-miles of coastal zones. On the contrary, *PlaNYC* proposes to intensify development wherever possible, in waterfront and other areas, without considering the risks posed by climate change. Finally, <u>PlaNYC</u> proposes no *ex-poste* strategy, or an emergency response to such disasters.

5.2.2.2 Material Measures

PlaNYC itself did not propose specific material measures for improving the adaptation of the city to environmental crisis and risk. There are other recent adaptation measures that the city is planning and undertaking mainly after the recent hurricanes that faced the city in recent years.

5.2.2.3 Urban Vulnerability Matrix

PlaNYC did not have in depth analysis for the threats and risk that face the city and its communities. It did not analyse the nature of the vulnerability demography in the city by age, gender, health, and other social or ethnic group. It also did not analyse the spatial distribution of risks, uncertainties, vulnerability and vulnerable communities in the city. The risk has been treated without any differentiations.

5.2.3 Equity and Justice of PlaNYC: Public Participation

New York is a diverse city with five boroughs, 59 community districts and hundreds of neighborhoods. *PlaNYC* acknowledges that shifting climate patterns will have a wide range of effects on these communities, taking lives and posing "major public health dangers," and impacting the property and livelihood of many (PlaNYC: 138). Moreover, all five New York City boroughs "have vulnerable coastline." Furthermore, the massive growth proposed by *PlaNYC* will certainly affect these communities, and may even "erase the character of communities across the city" (PlaNYC: 18). In considering the spatial impact of implementing the plan, the authors raise a crucial dilemma for the future of New York City and its communities:

> We cannot simply create as much capacity as possible; we must carefully consider the kind of city we want to become. We must ask which neighborhoods would suffer from the additional density and which ones would mature with an infusion of people, jobs, stores and transit. We must weigh the consequences of carbon emissions, air quality, and energy efficiency when we decide the patterns that will shape our city over the coming decades (PlaNYC: 18).

Despite the significant planning it embodies and the crucial dilemmas it raises, *PlaNYC* suggests no mechanism or procedure for facilitating citizen participation in the planning process, and makes no mention of public participation in the City's communities and neighborhoods. In short, careful reading of *PlaNYC* reveals markedly inadequate public participation in the planning process. *PlaNYC* asks: "What kind of city should we become?" and asserts: "We posed that question to New York" (PlaNYC: 9). However, instead of a systematic procedure for public participation in central planning, the planners employed participation methods that were disorganized at best:

Over the past three months, we have received thousands of ideas sent by email through our website; we've heard from over a thousand citizens, community leaders and advocates who came to our meetings to express their opinions; we have met with over 100 advocates and community organizations, held 11 Town Hall meetings, and delivered presentations around the city. The input we received suggested new ideas for consideration, shaped our thinking, reordered our priorities" (PlaNYC: 9).

Notwithstanding this process, it is clear that public participation in the process was inadequate and insufficient for meeting the planning challenges stemming from climate change for one of the world's most socially and culturally diverse cities. *PlaNYC* poses important urban dilemmas but does little to elicit real community participation. Instead, the planners appear to provide the answers themselves, in the name of New Yorkers: "By moving ahead, we will continue to ensure that the essential character of the city's communities remains intact as we seek out ... opportunities for public rezonings" (PlaNYC: 21).

Affordable housing appears to be one of the only themes that *PlaNYC* seeks to address. "The most pressing issue we face today is affordability," planners write. "Between 2002 and 2005 the number of apartments affordable to low-and moderate-income New Yorkers shrank by 205,000 units" (PlaNYC: 18). The Plan assumes that "if supply is not created as fast as people arrive, affordability could suffer further" (PlaNYC: 18). On this basis, it calls for expanding the housing "supply potential by 300,000–500,000 units to drive down the price of land" and for pairing "these actions with targeted affordability strategies like creative financing, expanding the use of inclusionary zoning, and developing homeownership programs for low-income New Yorkers." This, planners hold, will "ensure that new housing production matches our vision of New York as a city of opportunity for all" (PlaNYC: 12). However, what *PlaNYC* does in practice is to propose the provision of 500,000 housing units without proposing effective policies for ensuring affordable housing and regaining the more than 200,000 units that have already been lost. Moreover, *PlaNYC's* rezonings also have failed to adequately protect affordable housing (Paul 2010). Paul (2010: 4) proposes "that The affordable units created by the city's inclusionary zoning program (commonly known as the "80–20" because developers receive a subsidy for allocating 20 % of units to affordable housing) are outweighed by the loss of previously existing affordable units as market rents rise and rent-regulated tenants are pushed out by aggressive new landlords."

Although *PlaNYC* notes the existence of environmental injustice in the city, it fails to address the issue in a serious manner and takes no practical measures to mitigate the phenomenon. For example, planners acknowledge, most brownfields are concentrated in low-income communities, resulting in a case of severe environmental injustice (PlaNYC: 41). The owners of such land "often find that their financial interests dictate development plans that minimize cleanup requirements" and "may choose new uses for the land" that "do not reflect community needs or desires" (PlaNYC: 42–42). Moreover, "in some communities, the impacts of exposure to local air emissions have likely contributed to higher asthma rates and other diseases" (PlaNYC: 119). These clear cases of environmental injustices also go unaddressed by the plan.

PlaNYC encourages community involvement in significant planning issues *in the future* and reflects little interest in community involvement during the preparation of the plan itself. In this spirit, it suggests *future engagement* in developing adaptation strategies, mainly by working "with vulnerable neighborhoods to develop site-specific strategies," and to "create a community planning process to engage all stakeholders in community-specific climate adaptation strategies" (PlaNYC: 138). PlaNYC also suggests working with communities while exploring potential sites for development in their communities (PlaNYC, 25), and in the rezoning of brownfields (PlaNYC: 44). Angotti (2008a, b, c, d) criticizes *PlaNYC's* community participation and suggests that instead of carrying out a single discussion with each of the 59 Community Boards the plan "could challenge each community to come up with its own priorities for long-term sustainability, affordable housing creation, open space, and transportation. Let the communities speak."

Convincingly, Angotti (2010: 3) argues that the city's 59 community boards are:

> Still invisible in the 2030 plan barely mentioned in the scores of spreadsheets, maps and colorful images that herald the coming of the green city. They can post comments but play no role in setting priorities or initiating change. They are not consulted until after the fact, yet they are often criticized for only reacting. Civic and advocacy groups, including many that started fighting for a greener and greater future decades ago, and advocated sustainability long before the term was uttered in City Hall, are similarly sidelined.

Overall, *PlaNYC* focuses primarily on physical planning dimensions such as land, air, water, energy, and transportation in order to "unleash opportunity" (PlaNYC: 3) and less on socio-cultural issues. Virtually none of the major thrusts of the plan deal directly with issues of equity and justice, such as diversity, the future of communities and neighborhoods, poverty (which appears only once in the entire plan), and the cultural diversity of the city and its immigrants. Moreover, PlaNYC does not address the climate change vulnerability matrix, i.e., how climate change could affect each neighborhood, with an emphasis on the specific environmental risks that exist in each neighborhood and that each neighborhood is likely to face in the future.

The public participation process of *PlaNYC* was deficient as many planners and scholars argue. The New York Metro Chapter of the American Planning Association (NY Metro APA), which represents more than 1200 planners, designers, engineers and others involved in planning for NY metropolitan region (NYC, Long Island, and the Hudson Valley), criticizes *PlaNYC's* process and states: "The term 'sustainable development' has many different definitions and may be interpreted in a variety of ways depending on one's perspective, values or priorities. Therefore, credibility of the process and the plans depend on broad, open public involvement and accountability (NY Metro APA 2007: 2). Moreover, NY Metro APA (2007: 3) suggests that "particular attention should be given to seeking input from those least likely to have been reached by the extensive outreach to-date such as language minorities that reflect the diversity of New York, low-income residents, and those without ready access to the internet-based feedback mechanisms." Peter Marcuse (2008: 1) claims that *PlaNYC's* process of "participation was a sham." Angotti (2010: 1) contends that *PlaNYC 2030* left out any role for the city's hundreds of neighborhoods, 59 community boards, and the countless civic, community and environmental groups that care about the future of the city. And, that the plan was a top-down plan, conceived at City Hall with minimal input, and it was never approved as an official plan. Paul (2011: 1) claims that "the planning process for being too top-down and technocratic" with little community involvement. He argues that "in reality, public participation in *PlaNYC 2030* was an afterthought that was initiated only when the Mayor's office realized it was a necessary component of selling the plan to the public" (Brian 2011: 2). Rosan (2012: 973) concludes that "*PlaNYC* makes a number of important first steps for promoting EJ [environmental justice] through sustainability planning," and that "*PlaNYC* proposes a vast array of programs, many of which will improve the living conditions in EJ communities. Yet, there are instances where the EJ and activist communities feel excluded and ignored," and not all EJ or sustainability groups are satisfied with the plan or the process arguing that the plan fails to promote 'just sustainability'."

5.2.4 Urban Governance

PlaNYC advances an ambitious agenda for measures that aims to "create a sustainable New York City," which "will require tremendous effort: on the part of City officials and State legislators; by community leaders and our delegation in Washington; from the State government and from every New Yorker" (PlaNYC: 140). Nonetheless, planners acknowledge that "the existing organizations, programs, and processes are inadequate to implement these policies" and "no organization is currently empowered to develop a broad vision for energy planning in the city that considers supply and demand together as part of an integrated strategy" (PlaNYC: 104). The plan concludes that "there is a clear need for a more comprehensive, coordinated, and aggressive planning effort, focused on the specific needs of New York City," and therefore calls for the establishment of the New York City Energy Planning Board (PlaNYC: 105). It also calls for "changes at the City, State, and Federal levels—for transportation funding, for energy reform, for a national or state greenhouse gas policy" (PlaNYC: 11), and for "creating a new regional financing entity, the SMART Financing Authority, that will rely on three funding streams: the revenues from congestion pricing and an unprecedented commitment from New York City that we will ask New York State to match" (PlaNYC: 13). In addition, it suggests establishing a City office to promote brownfield planning and redevelopment (PlaNYC: 45). In the ways mentioned above, *PlaNYC* promotes an integrative approach to the issue of climate change on the formal institutional level. Nevertheless, it fails to effectively integrate civil society and grassroots organizations, such as the 59 Community Districts and the Boards of New York City.

5.2.5 Ecological Economics of PlaNYC

According to the authors of *PlaNYC*, improving the city's energy infrastructure and lowering demand will reduce energy costs by billions of dollars over the next decade; watershed protection will make multi-billion-dollar investment in new water filtration plants unnecessary; and improving public transportation and reducing congestion will reduce the economy's annual $13 billion loss due to traffic delays (PlaNYC: 133). By managing demand, increasing the energy supply, and saving energy in existing buildings, the city's overall power and heating bill will plunge by two to four billion dollars, resulting in an estimated annual savings of approximately $230 for the average household by 2015. Congestion pricing is projected to generate net revenues of $380 million in the first year of operation, increasing to over $900 million by 2030 (PlaNYC: 96). To this end, *PlaNYC* proposes an amendment to the City Charter requiring that New York City invest an amount equal to 10 % of its energy expenses in energy-saving measures each year. Planners also note that the measures required to execute these initiatives "will create thousands of well-paying jobs" (PlaNYC: 133), and that this will mean that the city will have "not only a healthier environment, but also a stronger economy" (PlaNYC: 13).

However, as we have seen, planners did not dedicate sufficient thought to solar energy in terms of design or as an alternative energy. PlaNYC suggests providing incentives to renewable energy and pilot emerging technologies, primarily for solar energy with the greatest potential. But the Plan also stipulates that "solar energy is still not as cost-effective as gas-fired electricity," and that New York City is uniquely expensive because taller buildings require more wires and cranes to carry equipment to rooftops, resulting in solar installation costs that are 30 % higher than in New Jersey and 50 % higher than in Long Island (PlaNYC: 112). In order to increase future solar use, the Plan suggests introducing property tax abatement for solar panel installations.

In these ways, *PlaNYC* provides a number of economic engines to promote climate change objectives and a cleaner environment. Its well based conclusion is that "adapting to climate change and investing in mitigation not only ensures the city's long-term economic vitality, but it will encourage public and private investments in the city's infrastructure, support green jobs, and improve the quality of life and level of service enjoyed by New Yorkers today." (*PlaNYC*: Progress Report 2009: 38).

5.2.6 Mitigation

Mitigation policies contain three main components: natural capital, energy, and eco-form as follows:

5.2.6.1 Natural Capital of PlaNYC

As we have seen, *PlaNYC* focuses on the dimensions of natural capital (air, water, and land) and proposes their efficient use in the future development of New York City. Its major strategies are to restore air quality, ensure clean water and waterfronts, collect runoff water, maximize land use and clean contaminated sites and brownfields, plant trees, and green the city. To this end, the Plan takes the following measures:

1. *Air*: Without action, the carbon emissions of New York City will grow to almost 74 million metric tons by 2030 (PlaNYC: 9). *PlaNYC* promotes initiatives to improve air quality and reduce emission by 30 % (PlaNYC: 116).
2. *Water*: The Plan calls for "developing critical backup systems for our aging water network to ensure long-term reliability" (PlaNYC: 12). It also proposes ways to maximize urban water absorption when planting trees (PlaNYC: 59). Finally, it suggests creating vegetated ditches (swales) along parkways to store direct rainfall and facilitate the natural cleansing of runoff (PlaNYC: 60).
3. *Waterfronts and Waterways*: New York City has 578 miles of waterfront, which the Plan regards as "one of the city's greatest opportunities for residential development," and an important site of other types of projects as well (PlaNYC: 22). *PlaNYC* also confronts the "legacy of the City's industrial past..." "...which treated New York's waterways as a delivery system" (PlaNYC: 51), and proposes to open 90 % of the City's waterways to recreation by preserving natural areas and reducing pollution (PlaNYC: 53).
4. *Trees*: "The City will expand efforts to reforest approximately 2000 acres of parkland by 2017," and reforestation will be implemented in many locations around the city (PlaNYC: 128).
5. *Land*: Since the City's land supply remains fixed, PlaNYC calls for using "our existing stock of land more efficiently" and recapturing almost all vacant, unutilized and under-used land for development.

5.2.6.2 Ecological Energy of PlaNYC

One major focus of *PlaNYC* is the city's energy sector. Its main aim in this realm is to provide cleaner, more reliable power for every New Yorker by upgrading the City's energy infrastructure (PlaNYC: 99). To this end, the plan calls for encouraging new cleaner power plants, renewing the city's most inefficient plants, and developing a market to increase the supply and use of renewable energy (PlaNYC: 103–115). In order to maximize energy efficiency, *PlaNYC* calls for focusing on buildings, the city's largest energy consumers (PlaNYC: 107). Over two thirds of the city's energy is consumed within buildings, compared to a national average of less than one third. According to the Plan, "the City has 5.2 billion square feet of space parceled into almost a million buildings" (PlaNYC: 107–108). By 2030, at least 85 % of the city's energy will be used by buildings that already exist today. In this way, energy saving measures in existing buildings will result in a seven million ton reduction in global warming emissions. This is significant, for without the measures outlined in the Plan, emissions would have risen to almost 80 million metric tons by 2030 (PlaNYC: Progress Report 2009: 39). *PlaNYC* also forecasts a 30 % reduction in the city's greenhouse gasses by 2030 (PlaNYC: 103).

In addition, the Plan proposes an extensive education and training campaign in the realm of energy awareness (PlaNYC: 110). It also encourages a shift to mass transit and various ways to promote fuel efficiency, the use of cleaner fuels, cleaner or upgraded engines, and the installation of anti-idling technology (PlaNYC: 13). According to the Plan, the most effective strategy is to reduce the number of vehicles on the road and to simultaneously expand the city transit system and implement congestion pricing (PlaNYC, 136). Planners predict that approximately 50 % of reductions in CO_2 emissions will come from increased energy efficiency in buildings, while 32 % will result from improved power generation and 18 % from changes in transportation. Planners explained their decision to not rely on "the widespread use of solar energy in this plan because its costs today are too high for general use" (PlaNYC: 136).

5.2.6.3 Eco-Form in PlaNYC

1. *Compactness*: Today, less than 4 % of the City's buildings account for roughly 50 % of the city's built area (PlaNYC: 102). *PlaNYC* proposes various planning strategies in order to increase compactness within the City. It suggests infill "everywhere it is possible" and development of spaces that "are now lightly used," such as parking lots in public housing areas that were developed in the 1930s (PlaNYC: 23). It also calls for developing underutilized areas throughout the city that are well-served by public transportation and other infrastructure; capturing the potential of transportation infrastructure investments; and decking over rail yards, rail lines, and highways (PlaNYC: 19–25). By *rezoning,* planners aim at "continuing to direct growth toward areas with strong transit access; reclaiming underused or inaccessible areas of our waterfront; and exploring opportunities to spur growth through the addition of transit, as our subways did more than a century ago" (PlaNYC: 21). *PlaNYC* fosters rezoning and redevelopment of brownfields, which according to the Plan represent one of the City's greatest opportunities, and which cover some 7600 acres throughout the five boroughs (PlaNYC: 41).
2. *Density*: New York is a dense city. Overall population density today stands at is 25,383 (persons per square mile), and the highest density in the city is 128,600 (New York City 2009). The planning strategies suggest further density intensification.
3. *Sustainable Transport*: The city's current transportation systems are in poor condition. More than half of the city's subway stations are awaiting repairs, and the city is more than $15 billion short of what it would cost to get the transit and road networks back into good shape. To make matters worse, trains are crowded, half of the subway routes experience congestion, and a large number of New Yorkers have no access to mass transit; (PlaNYC: 76). *PlaNYC* proposes a

"sweeping transportation plan" to enable the city to meet its needs through 2030 and beyond. The plan includes strategies to improve the transit network through major infrastructure expansion; improved bus service; an expanded ferry system and the completion of a master bike plan; and reduction of the increasing gridlock on the roads through better road management and congestion pricing (PlaNYC: 13). In addition, *PlaNYC* pursues transit-oriented development and uses rezoning to direct growth toward areas with strong transit access (PlaNYC: 21). As a result of these policies, New Yorkers will experience more comfortable travel, reduced travel times, and greater reliability, thus achieving a new standard of mobility (PlaNYC: 97).

4. *Mixed Land Uses*: *PlaNYC* encourages mixed land uses in future development, mainly by mixing transportation uses with residential areas and open spaces. Moreover, the plan encourages co-location of the 43,000 acres of city-owned land with other uses. Most of this land is developed for government operations, "but significant opportunities exist for housing to co-exist with the current use—from libraries to schools to parking lots" (PlaNYC: 22).

5. *Diversity*: *PlaNYC* recognizes that "the mixture of residents will determine, more than anything else, the kind of city we become," and that "by expanding supply possibilities to create healthier market conditions, we can continue ensuring that new housing production matches our vision of New York as a city of opportunity for all." "If New York loses its socioeconomic diversity," planners warn, "its greatest asset will be lost. We can—and must—do better" (PlaNYC: 27). On a practical level, however, *PlaNYC* neglects issues of socioeconomic and cultural diversity, including crucial socio-spatial issues such as segregation. It also fails to promote a wider variety of housing types.

6. *Passive Solar Design*: Although PlaNYC does not pay significant attention to passive solar design, it does suggest "greening" the Building Code of New York, with an emphasis on implementing the city's energy efficiency strategies, streamlining the process for incorporating new sustainable technologies into construction, and adapting to climate change. It also proposes focusing on reducing the amount of cement used in concrete, as cement production is an energy-intensive process that releases one ton of CO_2 for every ton of cement produced (PlaNYC: 106–107).

7. *Greening*: In New York City today, the standard park area per thousand residents is 1.5 acres, and there is an average of one playground for every 1250 children. Furthermore, in 97 out of the City's 188 neighbourhoods, the number of children per playground is higher (PlaNYC: 30). In this context, PlaNYC adopts greening as a major strategy and proposes three primary ways to ensure that by 2030, nearly every New Yorker will live no more than a 10-min walk from a park: (1) by upgrading land already designated as play space or parkland and making it available to new populations; (2) by expanding usable hours at current, high-quality sites; and (3) by re-conceptualizing streets and sidewalks as public spaces. The combined impact of these policies will be the creation of over 800 acres of upgraded parkland and open space across the city (PlaNYC: 31). *PlaNYC* also calls for beautifying the public realm and undertaking "an aggressive campaign to

plant trees wherever possible, in order to fully capitalize on tree opportunities across the city" (PlaNYC: 38). In addition, planners call for the expansion of "Greenstreets," a program that since its inception in 1996 has successfully transformed thousands of acres of unused road space into green space (PlaNYC: 38). They also suggest offering incentives for green roofs, which can reduce the volume of runoff by either absorbing or storing water and aiding other natural processes (PlaNYC: 60). Since the launch of *PlaNYC,* 200,000 trees have been planted across the five boroughs (PlaNYC: Progress Report 2009: 3).

8. *Renewal and Utilization*: Across the City, there are dozens of sites that are no longer suitable for their original intended use. *PlaNYC* proposes adapting unused schools, hospitals, and other outdated municipal sites for productive use as new housing (PlaNYC: 23). It also calls for cleaning and utilizing as many as 7600 acres of contaminated brownfields across the city (PlaNYC: 41) and suggests strategies to "make existing brownfield programs faster and more efficient; to create remediation guidelines for New York City cleanups; and to establish a City office to promote brownfield planning and redevelopment" (PlaNYC: 44). And, as we have seen, it calls for cleaning the water supply system and opening New York waterways for the use of residents (PlaNYC: 51–69).

9. *Scale*: *PlaNYC* focuses on plans for the city, streets, vacant and underused sites, buildings, and roof levels, but almost completely overlooks another important planning scale: the neighborhood. Angotti (2008a, b, c, d) states:

> But the other missing element is the region, including three states, hundreds of munici-palities and thousands of independent authorities. It may be mind-boggling, but how can you reduce the city's carbon footprint when the city is a fraction of the region's population and land? How can the city reduce traffic when Jersey and Connecticut are widening freeways? What about the city's waste being burned in the incinerator on the Jersey side? The task of even starting a regional discussion is daunting, but NYC is the most important player and the one most likely to kick it off.

In summary, as shown in Table 5.2 evaluating *PlaNYC* from the perspective of Eco-Form reveals that the plan actively promotes compactness and density; enhance mixed land uses; sustainable transportation; greening; and renewal and utilization. Its shortcomings are in passive solar design and planning for diversity.

Design concepts (criteria)	New York plan
Density	1. Low 2. Moderate 3. **High**
Diversity	1. **Low** 2. Moderate 3. High
Mixed land use	1. Low 2. Moderate 3. **High**
Compactness	1. Low 2. Moderate 3. **High**
Sustainable transportation	1. Low 2. Moderate 3. **High**
Passive solar design	1. **Low** 2. Moderate 3. High
Greening	1. Low 2. **Moderate** 3. High
Renewal and utilization	1. Low 2. **Moderate** 3. High
Planning scale	1. **Low** 2. Moderate 3. High
Total score	

Table 5.2 Eco-form matrix PlaNYC

5.3 Conclusions and Advice to Planners and Policy Makers

Using the proposed conceptual framework to evaluate *PlaNYC 2030* reveals important merits and shortcomings of the Plan. On the bright side, the Plan promotes greater compactness and density, enhanced mixed land use, sustainable transportation, greening, and renewal, and utilization of underused land. Finally, the Plan creates mechanisms to promote its climate change goals and to create a cleaner environment for economic investment, offers an ambitious vision of reducing emissions by 30 % and of a "greener, greater New York," and links this vision to the international agenda on climate change. On the down side, the assessment reveals that *PlaNYC* did not make a radical shift toward planning for climate change and adaptation and inadequately addresses social planning issues that are crucial to New York City. Like other cities, New York is "socially differentiated" in terms of communities' capacity to address the uncertainties of climate change, and the Plan fails to address issues facing vulnerable communities. Moreover, the Plan calls for an integrative approach to meeting the challenges of climate change on the institutional level but fails to effectively integrate civil society, communities, and grassroots organizations into the process. Another critical shortcoming, particularly during the current age of climate change uncertainty, is the lack of a systematic procedure for public participation in the planning process throughout the city's neighborhoods and among different social groupings and stakeholders. Practically, the proposed evaluation framework appears to be an effective and constructive means of illuminating the strengths and weaknesses of urban plans. It is also an easy-to-grasp evaluation method that can be easily understood and applied by scholars, practitioners, and policy makers. In light of the current climate change uncertainties, planners must profoundly rethink and revise the procedures and scope of conventional approaches to planning.

Based on the above evaluation of *PlaNYC* 2030, this chapter offers the following conclusions:

1. Like other cities around the world, New York's human, ecological, economic, and urban structures and spaces are at risk and face an increasing level of uncertainty due to the shifting parameters of climate change.
2. In light of these uncertainties, there is a need to rethink and revise the concepts, procedures, and scope of conventional approaches to planning. In order to meet the challenges posed by climate change, planning is in need of a more coordinated, holistic, and multidisciplinary approach, as planning in the context of such great uncertainty is unprecedented in our modern history.
3. Using the proposed conceptual framework to evaluate New York City's *PlaNYC 2030* provides an informative, easy to grasp, effective, and constructive means of illuminating the Plan's strengths and weaknesses.

4. Planning Approach:

   (a) *PlaNYC* is a strategic plan that aims to counter climate change. Significantly, climate change played a central role in formulating the plan's problem, justification, and visioning and objective settings. The Plan's outcomes were also climate change directed.
   (b) *PlaNYC* is a physically oriented plan, which mainly focuses on reconstructing infrastructures, promoting greater compactness and density, enhancing mixed land use, sustainable transportation, greening, and renewal and utilization of empty parcels and brownfields.
   (c) *PlaNYC* applies an integrated planning approach, making use of the advantages of new urbanism, TOD, sustainable development, mitigation, and monitoring institutional policies. It recommends efficient ways of using the city's natural capital assets and pays special attention to strategies for providing New York with cleaner and more reliable power. It creates a number of mechanisms to promote its climate change goals and to create a cleaner environment for economic investment.

5. The assessment reveals some of the merits of *PlaNYC*. It proposes effective measures for planning the physical dimensions of the city. In terms of eco-form, it promotes greater compactness and density, enhanced mixed land use, sustainable transportation, greening, and renewal and utilization. With regard to the concept of uncertainty, it addresses future uncertainties related to climate change with institutional measures, and enhances the urban adaptive planning capacity of the city. *PlaNYC* recommends efficient ways of using the city's natural capital assets and pays special attention to strategies for providing New York with cleaner and more reliable power. From the perspective of ecological economics, the Plan creates a number of mechanisms to promote its climate change goals and to create a cleaner environment for economic investment. Finally, *PlaNYC* offers an ambitious vision of reducing emissions by 30 % and creating a "greener, greater New York," and links this vision with the international discourse and agenda on climate change and sustainability.

6. According to the assessment, *PlaNYC* has three major shortcomings. The first is its failure to adequately address the social planning issues that are crucial to New York City, the most diverse city in the world. *PlaNYC* does not effectively address issues of equity, such as social justice, diversity, race, and economic segregation. It also fails to address the issues facing vulnerable communities due to climate change. New York City is "socially differentiated" in terms of the capacity of communities to meet climate change uncertainties, physical and economic impacts, and environmental hazards.

7. The second shortcoming of *PlaNYC* relates to the plan's adaptation strategy, which focuses on emissions reduction alone and fails to prepare the city and its physical infrastructure for potential disasters caused by climate change shifting. Unfortunately, PlaNYC did not make a sufficiently radical shift toward planning for climate change and adaptation. This being the case, it seems clear that the

authors of *PlaNYC* have not taken the lessons of Hurricane Katrina as seriously as they should.

8. The Plan's third shortcoming is that although *PlaNYC* calls for an integrative approach to climate change on the institutional level, it fails to effectively integrate civil society, communities, and grassroots organizations into the process. The lack of a systematic procedure for public participation throughout the city's neighborhoods and among different social groupings and other stakeholders is a critical shortcoming, particularly during the current age of climate change uncertainty.

9. Another important lesson we can learn from applying the proposed evaluation framework to *PlaNYC* is that when planning for climate change, planners must not overlook any one of the eight concepts of assessment. The framework is not a mere collection of unrelated concepts. Rather they are interconnected, with each concept playing a specific role in the evaluation and influencing the others. Based on the measures advanced in *PlaNYC*, New York City could certainly be "greener," but in order to truly be "greater," planners must better incorporate its main treasures—socio-cultural diversity and the people of the city—into the planning process and into the Plan.

5.4 Advice to Planners for Planning to Counter Climate Change

1. Planning to Counter Climate Change (PCCC) is an emerging approach in planning contemporary cities aiming at countering climate change impacts, adapting cities to future uncertainties, and protecting residents from environmental hazards and risk. Climate change played a central role in formulating the problem, visioning and goal setting, and the outcomes. PCCC is holistic in scope and in the issues it should cover, and it is integrative and multidisciplinary in its planning approach. Significantly, climate change challenges urban planning by adding another dimension of uncertainty about future conditions (Rosenzweig and Solecki 2010b). In fact, "climate change brings further uncertainties due to the evolving nature of the climate system, its potential impacts on many aspects of urban life, and the untested effectiveness of adaptation strategies" (Rosenzweig and Solecki 2010a: 14).

2. This chapter suggests that New York City should considerably update its PlaNYC and integrate the lessons of Hurricane Sandy into its planning efforts of New York City. Greening the city and planting millions of trees is not enough to make cities more resilient in facing future climate change impacts. There are critical needs for adequate adaptation measures. Therefore, Planning to Counter Climate Change should develop adequate adaptation policies, and planners should give adaptation issues high priority. Planners must develop a better understanding of the risks that climate change poses to infrastructure, households, and communities (Jabareen

2012). To address these risks, planners have two types of uncertainty or adaptation management at their disposal: (1) Ex-ante management, or actions taken to reduce and/or prevent risky events; and (2) Ex-post management, or actions taken to recover losses after a hazardous event (Heltberg et al. 2009).

3. Planning to Counter Climate Change should apply an inclusive, appropriate, and effective approach to public participation. Moreover, Planning to Counter Climate Change must address the current and predicted needs of vulnerable communities. There are individuals and groups within all societies who are more at risk than others and lack the capacity to adapt to climate change (Schneider et al. 2007: 7–19). Demographic, health, and socio-economic variables affect the ability of individuals and urban communities to face and cope with environmental risk and future uncertainties. These variables affect the mitigation of risk, response and recovery from natural disasters (Jabareen 2012; Ojerio et al. 2010). As a result, socio-economically weak communities are more vulnerable to suffer negative impacts, including property loss, physical harm, and psychological distress (Ojerio et al. 2010; Fothergill and Peek 2004). Therefore, climate change impacts at the urban level require even more inclusionary public participation processes in order to ensure that future planning take into account significant issues such as environmental justice, vulnerability, and specific needs of individuals and communities.

4. At the present, there is no single planning approach to cope with climate change at the city level. Therefore, Planning to Counter Climate Change takes advantages and integrates various planning approaches that aim to achieve its comprehensive objectives, and it is therefore, able to utilize the advantages of new urbanism, TOD, sustainable development, mitigation, adaptation, evaluation, and monitoring policies.

In sum, planning has a strong role to play in countering the impacts of climate change at the city level, and thus, climate change issues should play a central part in making future urban planning decisions. Apparently, climate change and its resulting uncertainties challenge the concepts, procedures, and scope of conventional approaches to planning, and create a need to rethink and revise current planning methods. Therefore, planning should be oriented to deal with uncertainties rather than adapting the conventional planning approaches.

References

Angotti, T. (2008a). The past and future of sustainability June 9. In *Gotham Gazette: The place for New York policy and politics*. http://www.gothamgazette.com.

Angotti, T. (2008b). Is New York's sustainability plan sustainable?" Hunter College CCPD Sustainability Watch Working Paper. http://maxweber.hunter.cuny.edu/urban/resources/ccpd/Working1.pdf.

Angotti, T. (2008c). *Is New York's sustainability plan sustainable?* Paper presented to the joint conference of the Association of Collegiate Schools of Planning and Association of European Schools of Planning (ACSP/AESOP), Chicago.

Angotti, T. (2008d). *New York for sale: Community Planning confronts global real estate.* Cambridge, MA: The MIT Press.

Angotti, T. (2010). PlaNYC at three: Time to Include the neighborhoods. *Gotham Gazette: The place for New York policy and politics.* http://www.gothamgazette.com.

Fothergill, A., & Peek, L. (2004). Poverty and disasters in the United States: A review of recent sociological findings. *Natural Hazards, 32*(1), 89–110.

Gehrels, W. R., Kirby, J. R., Prokoph, A., Newnham, R. M., Achertberg, E. P., Evans, H., et al. (2005). Onsetof recent rapid sea-level rise in the western Atlantic Ocean. *Quaternary Science Reviews, 24*(18–19), 2083–2100.

Heltberg, R., Siegel, P. B., & Jorgensen, S. L. (2009). Addressing human vulnerability to climate change: Toward a 'no-regrets' approach. *Global Environmental Change, 19*(2009), 89–99.

Holgate, S. J., & Woodworth, P. L. (2004). Evidence for enhanced coastal sea level rise during the 1990s. *Geophysical Research Letters, 31*, 1–4.

Jabareen, Y. (2006). Sustainable urban forms: Their typologies, models, and concepts. *Journal of Planning Education and Research, 26*(1), 38–52.

Jabareen, Y. (2008). A new conceptual framework for sustainable development. *Environment, Development and Sustainability, 10*(2), 179–192.

Jabareen, Y. (2012). Planning the resilient city: Concepts and strategies for coping with climate change and environmental risk. *Cities, 31*, 220–229.

Kern, K., & Alber, G. (2008). Governing climate change in cities: Modes of urban climate governance in multi-level systems. In *Competitive Cities and Climate Change, OECD Conference Proceedings, Milan, Italy, 9–10 Oct 2008* (Chap. 8, pp. 171–196). Paris: OECD. http://www.oecd.org/dataoecd/54/63/42545036.pdf.

Marcuse, P. (2008). *PlaNYC is not a "Plan" and it is not for "NYC".* Available from http://www. hunter.cuny.edu/ccpd/repository/files/planyc-is-not-a-plan-and-it-is-not-for-nyc.pdf.

New York City. (2009). http://www.nyc.gov/html/dcp/html/neighbor/neigh.shtml.

NPCC—New York City Panel on Climate Change: Climate Risk Information (2009). Available at http://www.nyc.gov/html/om/pdf/2009/NPCC_CRI.pdf.

NPCC—New York City Panel on Climate Change: Climate Risk Information (2009), Pais, J., & Elliot, J. (2008). Places as recovery machines: Vulnerability and neighborhood change after major hurricanes. *Social Forces, 86*, 1415–1453.

NY Metro APA—The New York Metro Chapter of the American Planning Association (2007). *Response to the Bloomberg Administration's PlaNYC 2030 long term sustainability planning process and proposed goals.* http://www.nyplanning.org/docs/PlaNYC_2030_response_final_ 3-14-07.pdf.

NYS. (2013). *NYS2100 Commission: Recommendations to Improve the Strength and Resilience of the Empire State's Infrastructure.*

Ojerio, R., Moseley, C., Lynn, K., & Bania, N. (2010). Limited involvement of socially vulnerable populations in federal programs to mitigate wildfire risk in Arizona. *Natural Hazards Review, 12*(1), 28–36.

Paul, B. (2010). How 'Transit-Oriented Development' Will Put More New Yorkers in Cars. *Gotham Gazette: The place for New York policy and politics.* http://www.gothamgazette.com

Paul, B. (2011). *PlaNYC: A model of public participation or corporate marketing?* http://www. hunter.cuny.edu.

PlaNYC 2030 (2007) PlaNYC 2030: A greener, greater New York. In *The City of New York.* New York: PlaNYC.

PlaNYC 2030. (2014). http://www.nyc.gov/html/planyc/html/about/about.shtml.

PlaNYC: Inventory of New York City Greenhouse Gas Emission. (2009). *Mayor's office of long-term planning and sustainability.* New York: City Hall. www.nyc.gov/PlaNYC2030.

Priemus, H., & Rietveld, P. (2009). Climate change, flood risk and spatial planning. *Built Environment, 35*(4), 425–431.

Rosan, C. D. (2012). Can PlaNYC make New York City "greener and greater" for everyone?: Sustainability planning and the promise of environmental justice. *Local Environment, 17*(9), 959–976.

Rosenzweig, C., & Solecki, W. (2010a). Introduction to climate change adaptation in New York City: Building a risk management response. *Annals of the New York Academy of Sciences, 1196*, 13–17 (Issue: New York City Panel on Climate Change 2010 Report).

Rosenzweig, C., & Solecki, W. (2010b). New York city adaptation in context (Chap. 1). *Annals of the New York Academy of Sciences* (Issue: New York City Panel on Climate Change 2010 Report).

Rosenzweig, C., Solecki, W. D., Hammer, S. A., & Mehrotra, S. (2010). Cities lead the way in climate-change action. *Nature, 467*, 909–911.

Schneider, S. H., Semenov, S., Patwardhan, A., Burton, I., Magadza, C. H. D., Oppenheimer, M., et al. (2007). Assessing key vulnerabilities and the risk from climate change. Climate change 2007: Impacts, adaptation and vulnerability. In M. L. Parry, O. F. Canziani, J. P. Palutikof, P. J. van der Linden, & C. E. Hanson (Eds.), *Contribution of Working Group II to the fourth assessment report of the intergovernmental panel on climate change* (PP. 779–810). Cambridge, UK: Cambridge University Press.

Schwab, J. C. (2010). *Hazard mitigation: Integrating best practices into planning*. Planning Advisory Service Report Number 560. Chicago, IL: APA—American Planning Association.

Solecki, W. (2012). Urban environmental challenges and climate change action in New York City. *Environment and Urbanization, 24*, 557–573.

Solecki, W. (2014). Urban environmental challenges and climate change action in New York City. *Environment and Urbanization, 24*, 557–573.

Swart, R., Biesbroek, R., Binnerup, S., Carter,T. R., Cowan, C., Henrichs, T, et al. (2009). *Europe adapts to climate change: Comparing national adaptation strategies*. PEER Report No 1. Helsinki: Partnership for European Environmental Research. Vammalan Kirjapaino Oy, Sastamala. Available online: http://www.peer.eu/fileadmin/user_upload/publications/PEER_Report1.pdf.

US Census Bureau (2009). American Community Survey: 2009 Data Release. http://www.census.gov/acs/.

van Leeuwen, E., Koetse, M., Koomen, E., & Rietveld, P. (2009). *Spatial economic research on climate change and adaptation. A literature review. Knowledge for climate programme.* Utrecht University. Available at http://www.kennisvoorklimaat.nl/nl/25222685KVK_Nieuws.html?opage_id=25222957&location=17222180632169871,10314425,true,true.

Chapter 6
Planning Practices of the Risk City Around the World

6.1 Introduction

This risk city theory suggests that significant changes in risk perception will lead to different practices. This premise rests on the assessment that new risk enhances the perception of lack and the need for actions and practices to fill this gap. The emergence of knowledge regarding the threats posed by climate change has contributed to new practices at the city level around the world, as reflected in new types of urban plans and planning practices introduced in response to the risks faced by contemporary cities.

In recent years, many cities have been grappling with climate change using master, strategic, and action plans aimed at mitigating greenhouse gas emissions and adapting to the anticipated, albeit uncertain, impacts of climate change. The critical importance of these recently issued city plans stems from the fact that they are currently the only vehicle for facilitating synthesis, practical integration, and synergetic application in guiding, visioning, and directing the future growth and development of cities, while at the same time addressing the many different risks posed by climate change.

It is important to emphasize that cities highly matter in coping climate change. Both the international community and the climate-change related discourse of local and international environmental civil society look to cities to play a leading role in coping with climate change. This expectation is premised on three main factors. The first is the scale of our contemporary cities, which will become home to the vast majority of humanity in the coming decades. Whereas only 29 % of the earth's population lived in cities in 1950, the figure today has reached 51 %, and by 2050 an estimated 70 % of the global population (6.3 billion people) will live in urban areas (UNDESA 2011). The second is the fact that today's cities have become a major source of greenhouse gas emission and are responsible for more than 70 % of global energy-related carbon dioxide emissions (WRI/WBCSD 2014). The third is the phenomenal risk that climate change poses to city populations and their social,

© Springer Science+Business Media Dordrecht 2015
Y. Jabareen, *The Risk City*, Lecture Notes in Energy 29,
DOI 10.1007/978-94-017-9768-9_6

economic, ecological, and physical systems (IPCC 2014), impacting urban security and threatening the safety, the well-being, and the very existence of urban people (Barnett and Adger 2005; Leichenko 2011; Rosenzweig et al. 2011). Without a doubt, cities as territorial entities represent one of the most promising vehicles and scales for tackling the challenges of climate change today.

Consequently, many cities propose new climate change actions plans, which are the product of tremendous efforts to counter climate change risk at the city level. Their primary significance lies in the role they will play in shaping spatial, social, economic, and security-related aspects of city life in Europe and other parts of the world. By means of these climate change-oriented plans, many cities, especially in developed countries, are now grappling with climate change through a multitude of practices aimed at mitigating greenhouse emissions and adapting to the anticipated, albeit uncertain impact of climate change. Despite their great importance, however, analysts have yet to assess the nature and impact of these plans at the national or cross-national level, or their possible effect on the environment and society.

Despite the monumental significance of these plans, however, analysts have yet to assess their nature and impact at the national and cross-national levels and their possible effect on the environment and society. Thus far, assessments have gone no further than reports on the climate change-related activities of cities–such as ARUP for the C40 (2011) and Broto and Bulkeley (2013)–based on information not gleaned from city plans, pertaining only to general activities and experiments conducted at the city level.

Nonetheless, we currently lack both the empirical analytic foundation necessary to determine the scale of emissions reduction that cities could potentially achieve, and sufficient evidence regarding past progress indicating what emissions would or would not have been had mitigation measures not been undertaken (Kennedy et al. 2012). Another essential question is whether cities are contending with climate change in a suitable manner by adequately reducing their emissions and improving their readiness and adaptation measures to face the uncertainties and threats it presents. A critical component of any answer to this question–one which the literature has thus far overlooked–must be an assessment of overall city mitigation and adaptation policies as reflected in their master and strategic plans. Our fundamental premise is that urban plans possess an unrivaled potential to contend with the impacts of climate change.

Scholars appear to be in agreement regarding the critical importance of understanding how cities today are planning to effectively respond to the impact of climate change and to promote sustainable urban habitats and the transition to increased urban resilience (Bicknell et al. 2009; Bulkeley 2013; Romero-Lankao and Qin 2011; Rosenzweig et. al 2010; Vale and Campanella 2005). Yet, when it comes to cities, we are still in the process of setting research agendas and formulating questions, and it is the time to draw conclusions regarding the specific and general impact of the measures that have been implemented over the past twenty years. Moreover, despite the importance of the recently issued plans and the substantial public resources that have been invested in their formulation, we still know very little about them and have yet to begun studying them and assessing their impact.

Therefore, this chapter asks critical questions about the nature, vision, practices, and potential impact of recent climate change-oriented plans. What kinds of risks do they attempt to address, what types of practices do they institute, and what types of approaches do they apply? Do they adequately address the risks and uncertainties posed? How do they contribute to the worldwide effort to reduce greenhouse gas emissions? Are the cities in question following the correct course of action according to the relevant European and international conventions? In this chapter, we also consider whether the plans address or marginalize equity and social issues and reflect social agendas. Answering these questions will enable us to understand whether the cities of the world are contending with climate change risk and uncertainties in a responsible manner, or, alternatively, becoming death traps for their residents by failing to effectively meet the challenges of countering climate change.

In asking these questions, this chapter aims to begin filling the gap in the literature by employing the assessment methods of Planning for Countering Climate Change (PCCC), which, as we have seen in a previous chapter, is comprised of six major concepts: utopian vision; equity and justice; adaptation; mitigation; ecological economics; and urban governance.

For the sake of our analysis, we selected ten city plans from around the world. The composition of this sample was based on the premise that it should include large cities (all of the selected cities are large cities in terms of population, and eight are state capitals); cities that are both developed and developing in character; and cities that have recently issued and approved city plans for the coming decades. The chapter will present an analysis of the ten plans in accordance with the six PCCC concepts of evaluation. The plan for New York City will be presented extensively in the following chapter and will therefore not be addressed here, although some aspects of the plan itself will be referred to within the tables of this chapter.

6.1.1 The Utopian Vision

From the perspective of the risk city, our analysis of the visions of the plans is looking for risk perceptions that motivate the vision and the planning efforts. It appears that each one of the selected plan presents a long-term vision for the city, extending to the years 2020, 2030, and further into the future. Some cities based their vision primarily on the risks and uncertainties stemming from climate change, while others offered visions that address other threats, such as those related to growth and urban expansion. All, however, address the lack stemming from existing or future threats. The plans vary with regard to urbanization and growth, demographic pressures, poverty and employment, housing, socio-economic conditions, economic development, and climate change-related aspects.

According to the *Paris Climate Protection Plan*: *Plan to Combat Global Warming* (2007), "the City of Paris has committed itself to a 'factor 4' approach with the aim of reducing total emissions in the area it administers and from its own specific activities by 75 % of their 2004 level by 2050" (p. 9). This commitment was enacted into French law by the Energy Policy Programming and Orientation Act of July 2005. On this basis, the plan contends that "the next half-century will mark a profound change in our civilization," reflecting its aim to reduce the "risks inherent" in human activities. The plan also highlights the fact that the "City of Paris and Parisian housing are vulnerable to climate change, and the city administration has accordingly decided to embark on discussions about the necessary adjustment strategy that this entails." Heat waves constitute the main issue of vulnerability, followed by the risk of the flooding of the Seine River. This vision is manifested in various practices proposed with the aim of coping with these risks.

With its target year of 2031, *The London Plan*: *Spatial Development Strategy for Greater London* (2011) has taken several different incarnations over the past decade. When Boris Johnson first took office as mayor in 2008, his advisors strongly recommended that he completely replace the London Plan of 2004. Accordingly, Johnson announced a full review of the London Plan soon after taking office, resulting in the formal publication of a replacement plan in 2011. It is interesting to note that the word "mayor" appears 577 times throughout the plan, indicating that for Johnson, the plan is both a powerful political instrument for visioning the future of the city and a task of major importance. In the plan's forward, Mayor Johnson articulates his vision for the city as follows:

> My vision for London embraces two objectives. London must retain and build upon its world city status as one of three business centers of global reach... London must also be among the best cities in the world to live, whatever your age or background. We need enough homes, meeting a diversity of needs. The local and distinctive have to be treasured. Our neighborhoods must be places where people feel safe and are proud to belong (*London Plan*, 2011: 6).

The plan's overarching vision, as rearticulated following Johnson's vision, casts London as a city demonstrating environmental responsibility for the entire world in its contribution to the undertaking of international risk reduction through its planning practices. "*London*," it asserts, "*should: excel among global cities—expanding opportunities* for all its people and enterprises, *achieving the highest environmental standards and quality of life* and *leading the world* in its approach to tackling the urban challenges of the 21st century, particularly that of climate change" (emphasis in source; Chapter: 26). Based on this over-arching vision, the plan goes on to identify six more detailed objectives regarding the future of the city, aiming to ensure that London is:

1. "A city that meets the challenges of economic and population growth..."
2. "An internationally competitive and successful city..."
3. "A city of diverse, strong, secure and accessible neighborhoods..."
4. "A city that delights the senses..."
5. "A city that becomes a world leader in improving the environment..."

6. "A city where it is easy, safe and convenient for everyone to access jobs, opportunities, facilities…" (*London Plan*, 2011: 6)

In addition to the theme of environmental responsibility, *growth* is another major concept of London's planning vision, which calls for economic and spatial growth to strengthen the city's international competitiveness and help fill the lack stemming from the socio-economic and spatial threats it faces. The plan is based on the premise that, in light of its growing and diverse population and "growing and ever changing economy," which exists alongside persistent problems of poverty and deprivation, "the only prudent course is to plan for continued growth" (p. 28). Ultimately, in order to realize the mayor's vision for the city, growth and change will be managed "across all parts of London to ensure it takes place within the current boundaries of Greater London without: (1) encroaching on the Green Belt, or on London's protected open spaces; and (2) having unacceptable impacts on the environment" (p. 33).

Barcelona's vision also emphasizes international competition and environmental responsibility. In the latter realm, the city's *Energy, Climate Change and Air Quality Plan 2011–2020* (PECQ) asserts that the city is currently facing risks stemming from climate change and other environmental threats. The "city infrastructures are at risk given rising sea levels, fluctuations in the supply of drinking water and sea storms, whilst the population is subject to the combined effect of increasing global temperatures, the heat island effect, the consequent reduction in air quality, and heat waves" (p. 30). With this in mind, the plan, led by the City Council of Barcelona, aims to "position Barcelona in approximately 2020 as a highly competitive city," and to provide the public administration with strategic tools to improve citizens' health and to improve the health of the planet by increasing energy efficiency and reducing greenhouse gas emissions, as well as other local effect pollutants (2010: 9).

Approved in 2010, *Master Plan of Moscow 2025* advances a vision by which Moscow and its population will enjoy a standard of living similar to that in other major European capitals and by which the city, the capital of the Russian Federation, will undergo development as a global city integrated into the global economy. According to official statistics, as of January 1, 2009 the population of Moscow stood at 10,509,000. Moscow is home to about 7 % of the population of the Russian Federation and is responsible for 20 % of the gross domestic product. The *Master Plan of Moscow* points out the following "positive trends of demographic development: the growth of the birth rate, decreased mortality, increases in life expectancy." According to the plan, however, the threats it faces are related to growth and expansion, and its "problems are exacerbated in the development of transport, engineering and social infrastructure." It is presented as "a master plan of needs" and "necessity" that "will strictly secure available construction sites, allocate normatively motivated construction sites, and determine necessary volumes of municipal project construction to secure social guaranties, on the one hand, and to provide the municipal construction with budgetary financing on the other hand" (*Master Plan of Moscow 2025*; The Government of Moscow-City; Ludmila

Tkachenko 2013). In this plan, growth and expansion emerge as a major strategic component of planning Moscow's future. On July 1, 2012, the jurisdiction of Moscow was officially expanded from 107,000 to 255,000 ha, and approximately 250,000 people were added to its population of nearly 12 million Muscovites. The city was expanded through the annexation of territories to the south and the west, including 21 municipalities and two urban districts (see Fig. 6.1).

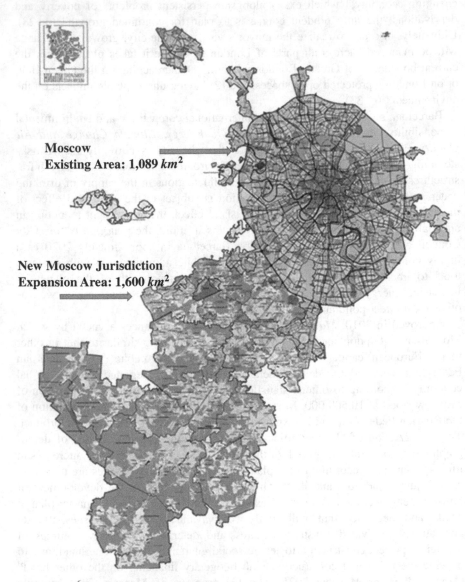

Fig. 6.1 The growth of Moscow according to its recent master plan

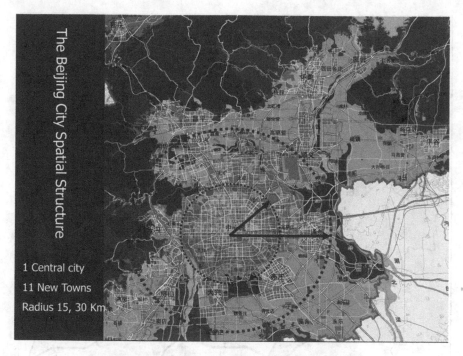

Fig. 6.2 Growth and new proposed cities according to Beijing master plan

The point of departure of *Beijing Master Plan, 2004–2020* (see Fig. 6.2), which was approved in January 2005 by China's State Council, is that "Beijing must take a broader view of the world to understand the dynamics of active development process" and therefore "must develop according to the standards of world cities and participate in the global order at a higher level" (*Beijing Master Plan*, p. 9). On the whole, the plan focuses on economic development and making the city more competitive in the global arena. To this end, it states that Beijing as a capital must promote "further development" and refers to "the 12th Five-Year Plan" as "vital for promoting the capital's scientific development." It also emphasizes the need to follow "the spirit of the Fifth Plenary Session of the 17th Party Central Committee and firmly grasp the main themes of scientific development" (p. 15). It is interesting to note that this particular plan reflects the modernist ideal of the early decades of the 20th Century, when states put their energies into "mastering nature" and promoting progress without taking into account environmental externalities and impacts. It is in this spirit that *Beijing Master Plan, 2004–2020* aims "to build Beijing into a World City" (p. 16) and to promote the overall "People's Beijing, High-Tech Beijing, and Green Beijing" (Part 1: 17). At the same time, the plan seeks to promote Beijing as an "internationally influential" city through the services it provides.

Like the abovementioned visions, the 2008 *Amman Plan: Metropolitan Growth* (see Fig. 6.3) also aims to accommodate Amman's projected growth until 2025.

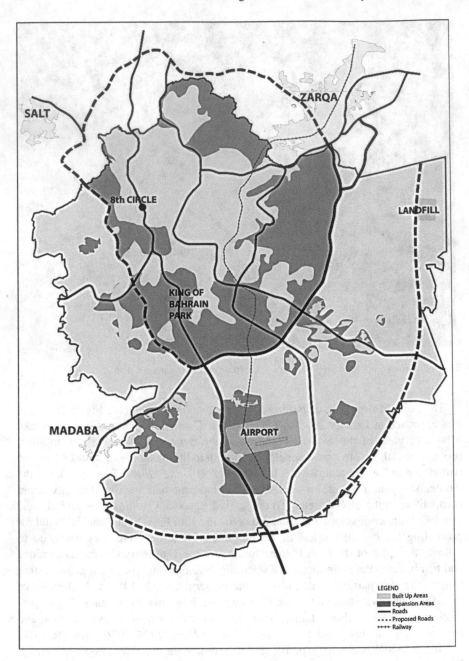

Fig. 6.3 The growth of Amman according to its recent master plan

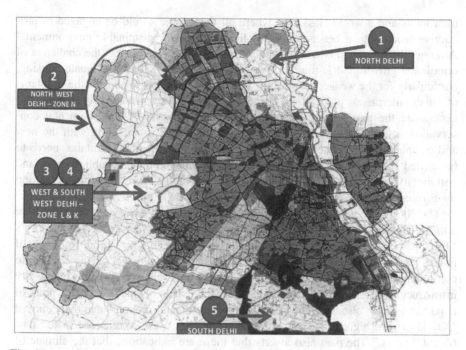

Fig. 6.4 Growth and new proposed cities according to MDP 2021 Delhi

Amman's vision is to achieve "an efficient city," "an inclusive and multicultural city," "a destination city for investment and visitors," and a city of heritage. Only at the bottom of the list does it seek to be "a green, sustainable city" and "a city for pedestrians." Without a doubt, the *Amman Plan* offers a vision that is modernist and economic growth-oriented and has a clear neoliberal oriented agenda. Indeed, the plan was initiated by the King of Jordan himself in an effort to attract investment, mainly from the Arab world, and to effectively channel the city's economic development.

The *Master Plan for Delhi* (MPD-2021) (see Fig. 6.4) depicts the urbanization processes currently facing the Indian capital as a major risk to the city in the present and the future. On this basis, the city's management of growth and urbanization are its primary concerns. The almost unprecedented pace and scale of urbanization in Delhi, which over the years has evolved into an conurbation, has resulted in enormous pressures on the physical environment and has had a severely adverse impact on pollution, making Delhi one of the most polluted cities in the world today. According to the plan, the experience of the city's past two master plans (1962 and 2001) reflects that although projections regarding various basic infra-structure services have been made with reference to population growth projections and increased urbanization requirements, the infrastructure provisions (especially those related to water and power) have not matched the pace of development. This renders particularly challenging the plan's 2021 vision of making Delhi "a global metropolis and a world-class city, where all the people would be engaged in productive work with a better quality of life, living in a sustainable environment." Among other things, this will require planning and action to meet the challenge of population growth and in-migration into Delhi, the provision of adequate housing, particularly for the weaker sections of society, and an effort to address the problems of small enterprises, particularly in the unorganized areas of the city. It will also necessitate the provision of adequate infrastructure services, environmental conservation, and the preservation of Delhi's heritage and its blending with the new and complex modern patterns of development. All this, the plan maintains, needs to be carried out within a framework of sustainable development, public-private and community participation, and a spirit of ownership and a sense of belonging among its citizens (MPD-2021, p. 17).

The 2009 *Guidelines for the Action Plan of the City of Sao Paulo for Mitigation and Adaptation to Climate Change*, which was prepared and approved by the city's Municipal Committee on Climate Change and Eco-economy, proposes a set of initiatives, constructed with the broad institutional involvement of the Sao Paulo community, that are aimed at responding to climate change and improving city infrastructure to raise the quality of life of the Paulistano people. The plan's point of departure is that "climate change is already scientifically proven," and that "current CO_2 levels are higher than any that the planet has ever experienced in the past 650,000 years." The plan also asserts that there are indications that the climate of the city of Sao Paulo has changed over at least the past seven decades. The analysis of the temporal series of maximum and minimum average annual temperatures recorded in the city reflects a trend toward higher temperatures. Especially evident is the temporal change in the average minimum temperature between 1933 and 2010, which rose from an average of about 13.2–15.4 °C (p. 15). The analysis also demonstrates a significant annual increase in total accumulated rain volume between 1933 and 2010. On this basis, the plan asserts, "it is important to properly address the types of risks faced by cities and to establish public policies for the mitigation, adaptation and management necessary to reduce their vulnerability" (p. 13). Based on "scientific" knowledge regarding the risk of climate change, the plan promotes practical strategies. For example, to contend with the risk stemming from climate change, it calls for promoting education and giving "priority to environmental health as a key resource in the development of life, by monitoring risk factors and implementing programs to control climate sensitive diseases" (p. 29).

Approved in 2010, the *Rome Climate Change Master Plan*, also known as *A Third Industrial Revolution: Master Plan to Transition Rome into the World's First Post-Carbon Biosphere City*, is one of the most ambitious plans recently produced in terms of its vision and practices. Its basic assumption, in the words of Jeremy Rifkin, one of its chief authors, is as follows:

> Never before in history has the human species found itself in such a precarious state. With our own survival on Earth now in question, an increasing number of scientists, government and business leaders, and civil society organizations are asking how to rethink urban life in

a way that will allow our species to flourish while ensuring the well-being of our fellow creatures and the ecosystems which sustain all life on the planet. (p. i)

In accordance with this premise, the plan aims to prepare Rome "to make the transition to a post-carbon Third Industrial Revolution economy between now and 2050 and to become the first city of the Biosphere Era" (p. i). The first plan of its kind, it "would remake Rome, embedding it within a surrounding biosphere park that would provide its inhabitants with a locally sustainable economic existence far into the future" (p. i). The utopian vision of Rome's plan is sweeping and far-reaching (p. 7):

In the 21st century, hundreds of millions of human beings will transform their buildings into power plants to harvest renewable energies on-site, store those energies in the form of hydrogen and share electricity with each other across continental inter-grids that act much like the Internet. The open source sharing of energy gives rise to collaborative energy spaces – not unlike the collaborative social spaces on the Internet.

The *Third Industrial Revolution* "not only organizes renewable energies, but also changes human consciousness (p. 1)." According to the plan,

We are in the early stages of a transformation to biosphere consciousness. When each of us is responsible for harnessing the Earth's renewable energy in the small swath of the bio-sphere where we dwell, but also realize that our survival and well-being depends on sharing our energy with each other across continental land masses, we come to see our inseparable ecological relationship to one another. We are beginning to understand that we are as deeply connected with one another in the ecosystems that make up the biosphere as we are in the social networks on the Internet.

Therefore,

Today, on the cusp of The Third Industrial Revolution, empathy is beginning to stretch beyond national boundaries to biosphere boundaries. We are coming to see the biosphere as our indivisible community and our fellow creatures as our extended evolutionary family... A truly global biosphere economy will require a global empathic embrace. We will need to think as a species–as *Homo empathicus*–and prepare the groundwork for an empathic civilization. (pp. 1–2)

Moreover, Rome's

Third Industrial Revolution vision will transform the agricultural region into a modern biosphere community: a place that can provide food for the industrial/commercial and historic/residential sectors, while preserving the local flora and fauna of the region for future generations. The agricultural region will be a living showcase of the Italian Slow Food Movement, combining state-of-the-art agricultural ecology and biodiversity practices. Open-air country markets, country inns and restaurants will feature local cuisine and promote the ecological and nutritional benefits of a Mediterranean diet. (p. 6)

Although not presented as part of this chapter, we note that the vision advanced by New York City in *PlaNYC* (which will be discussed extensively in the following chapter) calls for the city's development as a "greener, greater New York," and adequately addresses local and global climate change as a central concern of planning and future development.

Table 6.1 Visions of cities: orientations and concepts

City	Climate change risk-oriented vision		Vision's main concept
	Yes	No	
Paris	X		Local and global responsibility
New York	X		A "greener, greater New York." New York as a world city
Amman		X	Growth
Beijing		X	Beijing as a world city. Growth oriented
London	X		London as a world city and a global business center. Growth oriented. Environmental responsibility
Barcelona	X		Local and global responsibility
Moscow		X	Growth oriented
Delhi		X	Growth, housing, informality, squatters, and poverty. Traditional environmental terminology
Sao Paulo	X		Environmental concerns
Rome	X		Energy

To conclude, Table 6.1 denotes the main concept of the visions advanced of the ten city plans sampled. The plans can be divided into three basic categories with regard to their respective points of departure:

1. The first includes cities that acknowledge the threats to their cities and to humanity currently posed by climate change and proclaim their responsibility toward local places and regions, toward their cities, and toward the planet as a whole. The cities in this category—which include Paris, New York, Barcelona, Sao Paulo, and Rome—have articulated visions that are climate change risk-oriented in nature.
2. The second category includes cities whose major concern is growth and expansion and whose plans do not focus on climate change, despite the risks it poses to their growth and economic development. This category includes Beijing, Moscow, and Amman.
3. The third category includes cities whose concerns are primarily social (such as poverty and housing for the poor), environmental (in the traditional sense of the term, meaning, not regarding climate change but rather health and clean water), and physical (such as basic infrastructure and the establishment of new cities). In the sample, Delhi alone is representative of this type of city, which seeks to address issues of poverty, informality, squatters, and housing for the socio-economically underprivileged masses.

6.1.2 Adaptation

Adaptation, from the perspective of the risk city, is about 'fortifying' and securing the people, and the urban social, economic, physical, and environmental infra-structures, from any future threats and be prepared to various vulnerabilities and uncertainties. Our main question in this regard is how plans contribute to the adaptation policies and measures?

The concept of adaptation is comprised of three components: uncertainties, material measures, and the Urban Vulnerability Matrix. As shown in Table 6.2, none of the cities have taken adaptation measures seriously in their inclusive, master, and strategic city planning. In contrast, and with only slight differences, Paris, London, and New York have advanced limited adaptation measures, and none of the cities have adequately addressed the uncertainties relating to climate change and their anticipated local impacts. The plans considered also lack significant analysis of uncertainties and threats and uncertainty planning scenarios, and none of the cities have offered an Urban Vulnerability Matrix analysis. Barcelona's plan presents a spatial analysis of energy but fails to address various threats to specific neighborhoods and quarters.

London's plan (2011: 23) acknowledges that "some climate change is inevitable..." However "it is impossible" the plan explains, "to predict how these changes will impact on London. Specifically, it is likely that the direction and speed of change are such that the effects of this will be increasingly felt over the period of this London Plan." Moreover, the UK Government's latest climate change projections suggest that by the 2050s, London could see an increase of up to 2.7° in mean summer temperature, a 15 % increase in mean winter rainfall, and an 18 % decrease in mean summer rainfall over the 1961–1990 baseline. On this basis, the plan maintains that London has to be ready to deal with a warmer climate that is

Table 6.2 Uncertainty, adaptation, and the urban vulnerability matrix

City	Uncertainty		Adaptation measures		Urban vulnerability matrix	
	Yes	No	Yes	No	Yes	No
Paris	X		Limited			X
New York	X		Limited			X
Amman		X		X		X
Beijing		X		X		X
London	X		Limited			X
Barcelona	X			X		X
Moscow		X		X		X
Delhi		X		X		X
Sao Paulo	X			X		X
Rome	X			X		X

likely to be significantly wetter during the winter and drier during the summer. Therefore, adapting to the climate "will include making sure London is prepared for heat waves and their impacts, and addressing the consequence of the 'urban heat island' effect." The plan also recognizes that "heat impacts will have major implications for the quality of life in London, particularly for those with the fewest resources and living in accommodation least adapted to cope" (London Plan, 2011: 23). The city will also witness an increased probability of flooding, with higher sea levels. Higher and more frequent tidal surges, significant increases in peak flows of the Thames and other rivers, and the potential for more surface water flooding will also be an issue. A significant proportion of London's critical and emergency infrastructure is also likely to face increased risk from flooding, particularly if London experiences the kind of growth it anticipates by 2031. As it stands, there are already 1.5 million people and 480,000 properties in the floodplain. Another problem arising from climate change will be an increasing water shortage (London Plan, 2011: 24).

Although the development of the London Plan was "subject to a full Integrated Impact Assessment (IIA)," it applies conventional assessment approaches and pays insufficient attention to the differential risks and threats facing the city's communities and neighborhoods. The IIA satisfied the legal requirement of carrying out a Sustainability Appraisal (SA) (including a Strategic Environmental Assessment— SEA) and a Habitats Regulation Assessment (HRA). The IIA also included a Health Impact Assessment (HIA), an Equalities Impact Assessment (EqIA), and relevant aspects of the Community Safety Impact Assessment (CsIA) aimed at ensuring fulfillment of the statutory requirements of the Crime and Disorder Act of 1998 and the more recently enacted Police and Justice Act of 2006. The plan holds that adaptation policies should be "making sure buildings and the wider urban realm are designed with a changing climate in mind, encouraging urban greening—protecting, enhancing and expanding the city's stock of green space to help cool parts of the city, [and] continuing to manage and plan for flood risks" (p. 29). In this way, the London plan proposes limited and conventional adaptation measures and fails to address uncertainty scenarios.

Paris's plan calls for adaptation measures to cope with heat waves, which have been characterized as phenomena that "entail considerable risks in Paris" (p. 60). It also proposes adaptation measures for flooding, which "would affect 3,000,000 people, entail evacuating 270,000 and deprive 1,000,000 of electricity, heating and domestic hot water. It would handicap Paris' economy for months" (Paris Action Plan, 63).

The uncertainties identified by the Beijing plan are not related to climate change risk but to economic issues. It is in this context that the plan calls on the city to face the increasing external factors of uncertainty, "address this sense of anxiety… observe these uncertainties and soundly deal with them" (Beijing Master Plan: 13).

6.1.3 Equity—Justice and Planning Procedure

Equity, which represents a widespread spectrum of social issues including justice and rights to the city, is a fundamental concept in understanding the risk city and its practices. How much these practices and planning efforts contribute to critical urban issues of justice and rights.

Equity has to do with three major aspects of planning in general and of PCCC in particular: public participation in planning procedures, social aspects, and justice or fairness. Our analysis of the ten plans selected shows that in all plans (for developed and developing cities alike), public participation was extremely limited.

The Paris Action Plan had no direct public participation and did not involve the city's neighborhoods and communities in the planning process. Participation was limited to the plan's public presentation and to "sending posts and emails." The Bagatelle Gardens was the site of a public exhibition titled "*Energies Mode D'emploi: Le Paris du 21e Siècle*," which attracted 100,000 visitors and presented concrete solutions for tackling the climate change challenge (Paris Action Plan, 2007: 6). In the autumn of 2006, a section on climate change was added to the City of Paris's website, and the city later reported that "nearly 250 posts have since been received, with ideas on how to fight global warming" (Paris Action Plan, 2007: 6). In addition, "from June 2006 to January 2007, lecture-discussions were held in all the local ("*arrondissement*") town halls within Paris that so desired" for which "over a thousand people turned out to suggest and discuss proposals for combating global warming at a local level" (Paris Action Plan, 2007: 6). Comparable methods of "public participation" were employed in the formulation of New York's plan.

The planning approach of the Paris plan was based on GHG calculation. To this end, in 2004, the City of Paris launched a study to calculate the quantity of GHG being emitted by its own services and others in the Paris region (Paris Action Plan, 2007: 5–6). Planning was then based on the idea of conducting workshops with consultants, as reflected in workshops held on eight major areas with a direct impact on climate change: buildings, economic activity, transportation, consumption, cooperation, adapting to climate change, and education and awareness. The workshops were provided with emission quantification data for each sector, based on the 2006 carbon audit (*Bilan Carbone*) carried out by the city administration. Following these workshops, the city issued a White Paper on Climate, which sets out a shared vision of the areas for action. The Paris Climate Protection Plan reflects the city administration's commitment in response to the community's expression of views (Paris Action Plan, 2007: 5–7).

Although the London Plan's overarching vision notes issues of inequity (as indicated, for example, in the disparity in life expectancy across different parts of the city) as an overarching concern in its first chapter (Chap. 1: 14), its procedure lacked suitable public participation. A formal public comment process on the plan's development was conducted between October 2009 and January 2010, and responses were received from 944 authorities, developers, groups and individuals, for a total of 7166 separate comments. In addition, an independent panel appointed by the Secretary of State made 124 recommendations. This type of public participation process was "meant to reflect principles in the Aarhus Convention on access to information, public participation and access to justice in environmental matters, which was ratified by the UK Government" (Plan Intro, 11–12).

Despite the mayor's expressed desire to "see this Plan used by them [boroughs and neighborhoods] as a resource for localism, helping them develop and then implement local approaches to meet their needs, but also add up Londonwide to more than the sum of the parts" (p. 8), London's neighborhoods and boroughs were not involved in the planning process in a fitting manner. According to the London Plan, when assessing local communities' needs, particular attention should be paid to health and health inequalities, housing choice, mixed and balanced communities, social infrastructure, health and social care, education, sports facilities, improving opportunities for all, London's neighborhoods and communities, inclusive environments, and local open space.

With regard to housing supply decisions, the plan contains a call to implement a new approach that supports the mayor's "strategic responsibilities and priorities for managing and coordinating sustainable housing growth in London" and to recognize "the importance of housing supply to his economic, social and environmental priorities" and "London's status as a single housing market, while also taking a more bottom-up, participative and consensual approach" (Chap. 3, 10).

With regard to minorities and immigrants, the London Plan asserts that the city population has grown every year since 1988 and that its population will continue to grow until 2031. At this rate, by 2026, the city's 2011 population of 7.8 million is projected to increase to 8.57 million. More immigrants and people of childbearing age have moved to the city leading to strong natural population growth. London's population will also continue to diversify. According to the plan, "black, Asian and other minority ethnic communities are expected to grow strongly as a result of natural growth and continued migration from overseas" (p. 17). By 2031, an additional six London boroughs will likely have population majorities made up of these groups, with Harrow, Redbridge, Tower Hamlets, Ealing, Hounslow and Croydon joining Brent and Newham, which have been home to such majorities since 2001 (London Plan, 2011: 17). The plan also acknowledges that income poverty rates for children, working age adults, and pensioners are higher in London than elsewhere in the UK, with one-quarter of the city's working age adults and 41 % of all children living in poverty after housing costs are taken into account. "As a result, London is an increasingly polarized city" and "deprivation tends to be geographically concentrated" (London Plan, 2011: 22). These demographic figures support the critical professional and ethical assertion that the process of designing and formulating the plan should have involved minorities. In some cases, the plan calls for the engagement of residents, as in the case of regeneration: "The Mayor will expect regeneration programs to demonstrate active engagement with residents, businesses and other appropriate stakeholders" (p. 61). In terms of justice, the London Plan (2011: 73) maintains that "the Mayor is committed to ensuring equal life chances for all Londoners. Meeting the needs and expanding opportunities for all Londoners—and where appropriate, addressing the barriers to meeting the needs of particular groups and communities—is key to tackling the huge issue of inequality across London".

Similar to other cities, the Barcelona plan was also developed using a limited method of public participation through professional sessions. According to the plan, in order to secure the participation of the general public, of partners, and of sectors related to energy matters, a number of working sessions were proposed to define what the new plan needed to cover. This process, and the focus areas they identified, set the course that led to the development of the plan.

In the development of the Amman plan, public participation was extremely weak and the local communities played almost no part in the process. This lack of public participation appears particularly problematic when we consider that the plan intervenes on the community level and suggests planning modeling for the 228 existing neighborhoods in Amman, noting that "community plans for these neighborhoods will provide the highest level of planning detail, including detailed zoning and local road networks" (p. 25).

Beijing Master Plan, 2004–2020 represents an extreme top-down planning approach, as it is the central government, and specifically the Communist Party's Central Committee, who determined the nature of the planning and its vision, strategies, procedures, and outcomes. In this context, the Beijing Municipal Government is acting primarily as the "implementer" of the national government (Zhao 2011). Clearly reflecting the party's guiding role, the plan reads as follows:

We will consciously implement the spirit of the series of important instructions given by the Party Central Committee concerning Beijing's work, and take scientific development as our theme and transformation of our economic development patterns as the main line for our work. (Beijing Master Plan, Part 1: 17).

The Master Plan of Moscow, 2025 was also the product of an exceptionally top-down planning process and procedure which left no room for public participation. The Moscow City Duma enacted the Moscow city laws "On the Master Plan for Moscow" and "On the Rules of Land Tenure and Development in Moscow City." The Moscow City Government also dictated the regulations used by the Committee on Architecture and Urban Planning of Moscow City, which was responsible for the planning.

The *Master Plan of Delhi* (MDP-2021) cites "democratic procedure and statutory obligations" as reasons why "the Draft Plan was prepared after obtaining the views of the public" and why "it also included extensive consultations at the pre-planning stage by involving local bodies, Government of NCT of Delhi, public sector agencies, professional groups, resident welfare associations, elected representatives, etc." (p. 18). Unfortunately, the general public, the poor communities of the city, and its various social groups were not integrated into the planning process. Procedurally, the Ministry of Urban Development issued guidelines in 2003 for the preparation of MPD-2021, which inter alia emphasized the need to explore alternate methods of land assembly, private sector participation, and flexible land use and development norms. Objections and suggestions regarding the Draft Master Plan was solicited by Gazette notification on March 16, 2005 and public notification in newspapers on April 8, 2005. "In response, about 7000 objections/suggestions were received, which were considered by the Board of Enquiry which met on 17 occasions and also afforded personal hearing to about 611 persons/organizations." Ultimately, "the Ministry of Urban Development considered the proposal in the light of the inputs received…and finally approved the Master Plan for Delhi 2021 in the present form."

Nonetheless, the *Master Plan of Delhi* contains some significant social aspects. First of all, it determines that one of the most important aspects of planned development is the provision of adequate well-planned shelter and housing for the different categories of inhabitants of the city. The quantitative and qualitative shortages and deficiencies in this regard were observed in the course of formulation of MPD-2021 (p. 18). The plan identifies two major challenges it needs to address: "the phenomenon of unauthorized colonies and squatter/jhuggi jhompri settlements." "This reality," it asserts, "will have to be dealt with not only in its present manifestation, but also in terms of future growth and proliferation" (p. 18).

Like many of the other plans considered, the Rome plan also involved the use of workshops and the minimal involvement of the general public. Nonetheless, it also appears to offer some social advantages: namely, by enabling commercial businesses to reduce their energy bills and their carbon emissions in order to increase the value of their assets and to enjoy greater security against the long-term impact of rising energy prices. On a social level, as demonstrated by evidence across Europe, domestic energy efficiency can deliver significant reductions in household energy costs, particularly for low-income families.

In sum, as reflected in Table 6.3, the plans of Moscow and Beijing were formulated with almost no public participation, and public involvement in the formulation of the other plans was extremely limited.

6.1.4 Urban Governance

There is an assumption of the risk city framework that the emergence of new dramatic risks will cause cities and city planning to restructure their urban governance in an effort to better cope with the resulting threats and uncertainties, to effectively integrate the many involved stockholders, and to ultimately implement the proposed planning policies and practices. Yet, as Table 6.3 demonstrates, the selected plans reflect almost no shift in urban governance and the manner in which the cities in question aim to deal with the emerging risk. The only cities displaying slight changes are Paris, Sau Paulo, and New York City, all of which have proposed new agencies to deal with some aspects of the risk.

The Paris plan proposes establishing a local energy agency for the city, which will play a networking and monitoring role and work to pool public and private-sector resources and to provide technical support. In addition, this agency will encourage active involvement on the part of the City of Paris, Ile-de-France regional authorities, the French Environment and Energy Management Agency

Table 6.3 Public participation in city planning

City	Public participation			Shift in urban governance	
	None	Limited	Reasonable	Yes	No
Paris		X		X	
New York		X		X	
Amman		X			X
Beijing	X				X
London		X			X
Barcelona		X			X
Moscow	X				X
Delhi		X			X
Sao Paulo		X		X	
Rome		X			X

(ADEME), and other communities interested in taking part in the project through inter-communal cooperation (pp. 66–67).

The Beijing plan, on the other hand, reflects little concern regarding climate change issues. According to Zhao (2011: 27), "Beijing is leading energy intensity reduction efforts in China, but it does not yet have specific climate actions and policies." It is the top-down governance system that ensures that painfully little pressure is exerted on local governments to adopt local climate change mitigation policies ahead of national climate policies. At the present, most local city initiatives focus on energy efficiency improvement, energy structure change, and renewable energy development.

Based on the premise that information and knowledge regarding the impacts of climate change is critical, Sau Paulo's plan calls for improvements in the field of knowledge and information dissemination. To this end, the plan calls for generating climate change scenarios with high space-time resolution for the various watersheds in the municipality of Sao Paulo, for sharing data among all the observational platforms in the region, and for developing multimedia communication processes to publicize warning information about air quality, floods, and the potential for disaster with the speed and effectiveness required to make decisions necessary to protect life and material and financial goods. To support the implementation of the approved policy, city legislation that provided the plan's legal foundation established a collegiate and consultative body known as the Municipal Committee on Climate Change and Eco-economy, consisting of representatives of municipal and state government, civil society, popular entities, and the business and academic sectors. The results of the voluntary collaboration of these working groups are presented here as Guidelines for the City of Sao Paulo Action Plan for Mitigation and Adaptation to Climate Change, which were approved by the Municipal Committee on Climate Change and Eco-Economy.

6.1.5 Ecological Economics

In capitalist economies, the risk city supposed to use the "green economy," or ecological economics, for channeling practices of reducing risk in general, and threats stemming from climate change in particular at contemporary times. Yet, few plans and cities promote this strategy adequately. Some neoliberal advocates argue that this undertaking should be left to the market itself and that planning should not intervene. As reflected in Table 6.4, there are variations among cities vis-à-vis their approach to the green economy.

The Paris Action Plan calls for promoting a partnership between the government and private sectors to "foster economic activity and boost job creation concerned with the fight against climate change."

Amman's plan demonstrates no green economic approach whatsoever, clearly reflecting the fact that its major goal was to use zoning to facilitate economic development. Beijing's plan is also geared toward economic development and demonstrates no concern for the principles of ecological economics, as reflected in its assertion that "we will be first to form a development pattern driven by innovation, increasing our comprehensive economic strength by strengthening our competitiveness significantly and by improving functions that serve national development" (Beijing Master Plan, Part I: 21). In a similar spirit, it promises that "industries will be upgraded and consumption and investment will be strengthened to boost the Capital's economic development to a better, faster and higher level" (Part II: 41).

Unlike these two plans, Sao Paulo's plan sets guidelines for the development of a low-carbon urban economy and calls for studying forms of payment for environmental services offered by the preservation of natural resources and the viability of the creation of economic and fiscal incentives for the use of renewable energy sources. This requires the development of economic and fiscal incentives for sustainability and the creation of funds for mitigation and adaptation to climate changes (p. 29).

Table 6.4 The ecological economics of the city plans

City	Ecological economics			
	Virtually none	Low level	Medium level	High level
Paris		X		
New York			X	
Amman	X			
Beijing	X			
London		X		
Barcelona			X	
Moscow	X			
Sao Paulo			X	
Rome				X

According to Rome's plan (p. 12), the costs of climate mitigation within Europe, which is moving toward the equivalent of a Third Industrial Revolution, could require as much as 0.5 % of the GDP by 2030. For Rome, it appears that the investment magnitude could be somewhat lower, closer to 0.3 % of the GDP. At the same time, improving energy efficiency has the potential to reduce the cost of living in Rome, thereby releasing substantial resources back into the local economy for other productive investment. At current energy prices, if Rome were to achieve its target of a 20 % improvement in energy efficiency per unit of GDP, the city would save €800 million per year (expressed in constant 2008 Euros). Assuming that these savings were consumed or invested in line with current economic patterns, the energy savings can be expected to generate an additional €230 million of economic growth per annum.

6.1.6 Mitigation

While adaptation is about securing the risk city, mitigation is about contribution to the preventing of occurring risks and threats. Mitigation has the three main components that are related to prevention: natural capital, eco-form, and energy.

6.1.6.1 Natural Capital—The Level of GHG Reduction

In this context, natural capital focuses on air quality and levels of GHG reduction, which ranges from 0 to 70 %. The target of the Paris Action Plan is "very ambitious," aiming for a 25 % reduction in greenhouse gas emissions in the city (in comparison to European targets of 20 %) by 2020. The London Plan strives for a 60 % reduction in London's overall carbon dioxide emissions (to below 1990 levels) by 2025 (p. 140). The London Plan also acknowledges the fact that the UK is the world's eighth largest emitter of carbon dioxide and that London is responsible for 8.4 % of these emissions (44.71 million tonnes, according to the most recent annual estimate). In 2008, Barcelona signed the European Union's Covenant of Mayors, committing to reduce CO_2 emissions by 20 %, to increase energy efficiency by 20 %, and to ensure that 20 % of its energy will come from renewable sources—all by 2020.

Unlike the Paris, London, and Barcelona plans, the plans for Beijing, Delhi, and Amman provide no figures regarding emission reduction. The Beijing plan states only that "Beijing will make even greater efforts to cope with pollution and reduce polluting emissions" and that the Clean Air Action Program of Beijing will be completely implemented (Part VI: 185).

The plan for Delhi also offers no target figure for GHG reduction and only acknowledges that "the air quality has been responsible for a number of respiratory diseases, heart ailments, eye irritation, asthma, etc." The two main sources of air pollution in Delhi are vehicular emission (approximately 70 %) and industrial emissions (approximately 20 %). In addition to pollution stemming from industries, the major area of planning and intervention relates to transportation planning. With the phenomenal growth in the number of vehicles in the city (representing an increase of 800–1000 % in absolute terms over the past two decades), the most significant aspect in the context of congestion and pollution has to do with the growth in personalized transport as compared to the availability of public transport. The enormous share of private vehicles in Delhi creates tremendous pressure on road space and parking, and also generates immense pollution, both directly and through congestion (p. 110).

6.1.6.2 Energy: The Energy Planning Turn

Although energy is a key concept of many recent plans around the world and a critical concept of mitigation, some cities have intensified their efforts in energy planning while others have dismissed it. Paris's plan advances a goal of a 25 % reduction in energy consumption based on the use of renewable energy sources, as well as a 30 % reduction in the energy consumption of municipal services and street lighting and a 30 % share of renewable energies in its energy mix. To this end, the plan sets overall targets for "low energy buildings" and suggests a maximum primary energy consumption (i.e., for heating, hot water, lighting, ventilation, and air conditioning) of 50 kWh/m^2 of net floor area per year for new-build operations, and 80 kWh/m^2 of net floor area per year for renovated buildings. In addition, it proposes conducting a thermal audit of its 3000 public facilities within three years and calls for launching a plan to renovate its building stock, including the thermal renovation of buildings, the renewal of heating and ventilation equipment, the reduction of electricity consumption, the improvement of street lighting management, and an increase in the share of renewable energy in consumption. The city also strives to promote the renovation of Paris's 100,000 buildings by 2050. Moreover, in conjunction with the Regional Council, the FFB (French Building Federation) for the Paris Île-de-France Region, CAPEB (the small builders' trade body for Paris and the inner suburbs), the Parisian federation of cooperative companies in the building and public works sector (*Fédération parisienne des SCOP du BTP*), and ANAH (National Housing Improvement Agency), the City of Paris has decided to draw up a partnership agreement to encourage Parisians to carry out specific, effective work to combat climate change. As an example, a system of energy-saving certificates has been set up to meet the objectives of reducing energy intensity laid down by law.

Beijing's plan proposes to "optimize our energy structure" and declares that "the use of clean energy supplies–such as natural gas–will be increased to a large extent; coal consumption will be reduced; smoke pollution will be strictly controlled" (Beijing Master Plan, Part VI: 186). The plan also undertakes to promote "consumption patterns to actively cope with climate change." Despite these declarations, however, the plan fails to adequately address these objectives in practice.

The London Plan proclaims that the city's mayor has formulated strategies for climate change adaptation, climate change mitigation, and energy management, as well as strategies related to waste management, air-quality, and water and biodiversity. The plan outlines specific policies such as retrofitting, the development of more decentralized energy networks, renewable energy, innovative energy technologies, adaptation to weather changes, urban greening (including green roofs), and the promotion of sustainable water and waste management. According to the plan (p. 138), London's greatest challenge "is to improve the contribution of the existing building stock (80 % of which will be still standing in 2050) to mitigating and adapting to climate change." To this end, the plan maintains that

> The Mayor will work with boroughs and developers to ensure that major developments meet the following targets for carbon dioxide emissions reduction in buildings. These targets are expressed as minimum improvements over the Target Emission Rate (TER) outlined in the national Building Regulations leading to zero carbon residential buildings from 2016 and *zero carbon non-domestic buildings* from 2019 (p. 141).

The London Plan also points out that "encouraging energy efficiency is important for reasons going beyond climate change" and that "a growing city with more households and jobs will need reliable and sustainable supplies of electricity and gas to power its homes, offices and other workplaces, transport network and leisure facilities." To this end, it seeks to increase the proportion of energy generated from renewable sources, and announces that the "Government has adopted a UK wide target for 15 % of total energy to be generated by renewable sources by 2020," and that "these projections represent London's contribution to this 2020 target and beyond" (p. 150).

Sao Paulo's plan calls for increasing the energy efficiency of buildings and electronic equipment, for stimulating the generation of renewable and decentralized energy, and for giving priority to the use of new energy sources. It also proposes the use of renewable energy sources through the development of projects to create incentives to adopt new sources and utilize energy from urban solid waste, by means of a viability study of new technologies (p. 41). In addition, the plan promotes and encourages standards of efficiency for the use of natural resources in new and existing buildings based on their adherence to the realities in the municipality and to the potential socio-environmental impacts (p. 45).

Rome's *Third Industrial Revolution Master Plan* is built on a foundation of increased energy efficiency and maximum utility from increasingly scarce resources. According to the plan, the four pillars of the Third Industrial Revolution are (p. 7):

1. The expanded generation and use of renewable energy resources—gathering the abundant energy available across our planet, wherever the sun shines, the wind blows, biomass and garbage are available, the tides wax and wane, or geothermal power exists beneath our feet.
2. The use of buildings as power plants—recognizing that homes, offices, schools and factories, which today consume vast quantities of carbon producing fossil fuels, could tomorrow become renewable energy power plants.

3. The development of hydrogen and other storage technologies—storing surplus energy to be released in the times when the sun isn't shining or the wind isn't blowing.
4. A shift to smart-grids and plug-in vehicles—the development of a new energy infrastructure and transport system that is both smart and agile.

The strategic objectives of Barcelona's plan are energy-oriented, and aim to "position Barcelona in the current context of energy at the level of Catalonia, Spain and Europe, and redefine its energy strategy with new objectives and action plans"; to "establish a municipal strategy with regard to climate change and air quality, fully coordinated with the energy strategy"; and to "raise awareness" and "generate a climate of involvement amongst all agents that participate in conceiving and executing the new Plan." To these ends, the plan aims to improve energy efficiency, reduce energy consumption in the city, cut the increase of greenhouse gas emissions (GHGs), and improve urban air quality and the quality of the energy supply. In 2002, the Barcelona City Council approved the PMEB (Barcelona Energy Improvement Plan) for 2001–2010, a municipal action plan involving a range of projects and measures aimed at increasing energy efficiency, reducing greenhouse gases, and increasing energy production using sustainable sources.

Barcelona's plan presents an innovative approach to planning the urban aspects of energy. Its methodology offers a number of new elements and themes and updates specific emission and energy efficiency factors. They include the spatial distribution of pollutants, the people's perspectives on energy use, the study of air quality in Barcelona (modelling the dispersion of pollutants and detecting their origin), the categorization of vehicles in the city and their pollutant emission levels, and an economic analysis of the plan's implementation in terms of new business opportunities, new jobs, and other such considerations. Other new elements include a study of the port's and airport's economic and environmental effects on the city, an updating of the emission factors of different pollutants in accordance with internationally applied methodologies, an analysis of industrial energy performance, a detailed study of the public buildings and facilities sector, and a study of safety in the supply of energy (p. 26). The Barcelona Plan also applies (p. 26) a number of instruments to conduct an in-depth study of energy performance in different sectors and the integration of all the results of the sectorial analyses into a single tool. The key applications employed were as follows:

(a) *Geographic Information System (GIS)*: A tool that makes it possible to link large databases with territorial coordinates to produce geo-referenced databases that can represent maps or facilitate territorial analysis.
(b) *A pollutant dispersion model* that makes it possible to analyze air quality based on an emissions inventory by means of pollutant dispersion modeling and chemical reactions that may be generated in the area under study.
(c) *An overall analysis model for the city* that makes it possible to carry out an analysis of the city and of the typologies that comprise it from the perspective of energy and greenhouse gases.

(d) *A classification tool for projects and grouping scenarios* that classifies, orders, and groups the plan's projects according to different criteria in order to define scenarios, make decisions about prioritization, visualize the environmental effects of the measures, and model the applications of the projects over time.
(e) *A thermal simulation model for buildings* that facilitates a dynamic analysis of the thermal demand and consumption of buildings (by means of building typologies).
(f) *An economic model* allowing for simulations of economic forecasts for Barcelona.
(g) A tool for detecting and analyzing vehicles in the city, which categorizes vehicles according to technical and environmental criteria using a system that reads license plates.

6.1.6.3 Eco-Form

As mentioned in the previous chapter, eco-form has several sub-components. Generally, it is about promoting and restructuring a transportation system that is more compact, dense, green, reusable, and sustainable. The analysis shows that the selected cities have demonstrated differing approaches to these concepts, as reflected in Table 6.5.

The Paris Climate Protection Plan includes measures recommended in the Paris Transport Plan, which promotes reductions in greenhouse gas emissions from Paris traffic in particular. In addition, in accordance with the goals of the SDRIF (*Schéma directeur de la région Ile de-France*—a regional development and urban-planning

Table 6.5 Mitigation measures: energy, GHG levels, and eco-form

City	Mitigation					
	Energy		GHG emissions goal		Eco-form	
	Renewable	No	% Reduction	Target year	Yes	No
Paris	25 %		70	2050	X	
			25	2020 (baseline 2004)		
New York	20 % (2020)		30	2050 (baseline 2005)	X	
	60 %					
Amman	0		0			X
Beijing	0		0			X
London	15 %		60	2025 (baseline 1990)	X	
Barcelona	20 %		23.45	2020 (baseline 2008)		
Moscow	4.5 % (2020)		0			X
Delhi	0		0			X
Sao Paulo	69 %		20	2020 (baseline 2005)	X	
Rome	20 %		20		X	

master plan for the Ile de-France region), the plan calls on the City of Paris to give precedence to dense development in the projects it undertakes and in which it is involved. Its aim in so doing is to contain urban sprawl, limit the city's ecological impact, and curb the use of private cars by providing a suitable alternative in the form of public transport and "soft" transport (Paris Action Plan, p. 37).

The Amman plan is a traditional land-use zoning plan consisting of a number of plans for "tall buildings," a "corridor intensification strategy," an "industrial lands policy," an "outlying settlements policy," an "airport corridor plan," and a "metropolitan growth plan." The Plan encourages compact urban growth in order to make the best possible use of existing services, promote increased transit use, improve pedestrian accessibility, and improve affordability for both the Greater Amman Municipality and its residents. At the same time, it overlooks the aims of sustainable modes of transportation and greening the city.

Beijing's plan reaches beyond the existing city borders and calls for building new cites in the countryside and around Beijing and for making use of agricultural lands and green lands for this purpose. According to the plan, the "focus of urban development will be shifted to new development districts," and "the constructions of new satellite cities and relatively underdeveloped areas will be accelerated" (Part V: 137). The plan all but neglects concepts of sustainable transportation and sustainable modes of spatial planning of eco-forms and instead proposes intensive development of the country's road network, similar to the post-WWII development of transportation networks in Europe and the US. The Moscow Plan foresees a substantial increase in the number of "personal automobiles" and therefore also calls for developing the roads.

To reduce CO_2 emissions by 20 % by 2020, increase energy efficiency by 20 %, and ensure that 20 % of the energy consumed comes from renewable sources, Barcelona's plan has two main programs: the Municipal Program and the City Program. Each has its own targets and strategies of energy saving, efficiency, emissions reduction, and the usage of renewable energies (which can break the perpetuation of some myths and barriers linked to certain attitudes and technologies). The Municipal Program targets buildings and resources that are under the direct responsibility of the city itself and works transversally together with other plans of the Barcelona City Council (such as the Urban Mobility Plan, the Green Spaces Plan, and the Tourism Plan), as well as with other municipal actors. All the elements considered within the Municipal Program (such as public buildings, lighting, municipal fleets, and urban services) consumed 473 GWh during 2008 and emitted 84,400 tonnes of greenhouse gas, or 2.8 % of the city's total energy consumption. These municipal consumption rates provide a baseline for achieving a 20 % reduction in related emissions by 2020, by means of the 23 projects which together constitute the Municipal Program. All city council stakeholders were given the opportunity to make proposals for the program and to express their opinion during the participation process that was conducted within the framework of planning.

According to *The Master Plan for Delhi*, the city contained an estimated built-up area of 702 km^2 in 2001, accommodating a population of approximately 13.8

million. To accommodate the population of 23 million projected for the year 2021, the plan recommends a three-pronged strategy consisting of encouraging deflection of the population to the NCR towns; increasing the population holding capacity of the area within the existing urban limits through redevelopment; and extending the present urban limits to the extent necessary.

On this basis, *The Master Plan for Delhi* asserts the necessity of redevelopment, densification, and city improvements in the existing urban areas. This major new aspect of the Master Plan will require a comprehensive redevelopment strategy for accommodating a larger population and for strengthening infrastructure facilities accompanied by the creation of more open spaces at the local level by undertaking measures for redevelopment of congested areas (p. 18). Another challenge observed by the Master Plan is the phenomenal growth of automobiles in Delhi. This has resulted in a variety of problems pertaining to congestion, pollution, travel safety, parking, and other issues that need to be addressed. In this way, the plan proposes the creation of a sustainable physical and social environment for improving quality of life as one of its major objectives (p. 107).

Delhi, including New Delhi (the NDMC area), contains a large number of residential, commercial, and industrial areas that are old and characterized by poor structural conditions, sub-optimal land, congestion, poor urban from, inadequate infrastructure services, lack of community facilities, etc. It is in this context that the plan contemplates a mechanism for restructuring the city based on mass transport. The Perspective Plan of physical infrastructure prepared by the relevant service agencies should help better coordinate and augment the services. The plan proposes a re-development and restructuring package based on: (1) incentivized redevelopment with additional FAR, which has been envisaged as a major element of city development covering all areas to encourage the growth impulse for regeneration in the target redevelopment areas (p. 43); (2) re-densification of low-density areas; and (3) the redevelopment of other developed areas.

Delhi's plan proposes to ensure "shelter for all," particularly for the vulnerable groups and the poor, by harnessing the potential of the public, the private/corporate sector, and the household sectors to create an adequate housing stock on either a rental or ownership basis (p. 52). Based on the projected population of 23 million by 2021, the city will require an estimated additional 2.4 million residential units. For housing the poor, the plan proposes the following solutions: on-site slum rehabilitation; measures to prevent slum expansion; housing for 50–55 % of the total urban poor; re-categorization of housing types and the development of control norms and differential densities to make housing viable and economical; and the normalization of unauthorized colonies, in accordance with government policy, and their effective incorporation into the mainstream of urban development. This will require infrastructure development and the provision of services and facilities, for which differential norms and procedures have already been devised.

According to Sao Paulo's plan, "giving priority to the use of collective public transportation, promoting a change in energy sources, and increasing the use of renewable fuels and clean energy are the main focuses indicated by the Transportation Working Group to improve climate conditions in Sao Paulo" (p. 25).

The plan gives priority to non-motorized transportation and to changing the energy matrix to renewable fuels and clean energy. It also promotes the development of a compact city, which seeks "equilibrium in the relationship between locations of employment and residence by means of urban interventions that stimulate the creation of new regional centers with different vocations, and re-qualify and revitalize existing ones, especially in conjunction with the high capacity transportation network, in order to guarantee greater transportation efficiency" (p. 49).

The urban form proposed by Rome's plan is unique. In terms of restructuring, the plan calls for transforming "now defunct commercial buildings into new residential blocks, using innovative architectural techniques that echo some of the best elements of ancient Roman building design." Moreover, "surrounding a newly revitalized residential city center will be the green industrial/commercial circle—the dynamic hub of Rome's economy, providing accessible jobs for the population." The plan also promotes the greening of Rome, which "will include thousands of small public gardens scattered in neighborhoods across the historic/residential core…Thousands of other small gardens will be placed in public areas around the city as part of the long term-plan to transform Rome into a sustainable biosphere park." (p. 4). It also suggests developing urban agriculture by making use of the city's 80,000 ha (of the total 150,000 ha which modern Rome currently occupies) of designated green space, which it regards as an under-used resource that could be made more agriculturally productive and serve as a site for leisure activities and large-scale renewable energy generation.

6.2 Conclusions

Based on the above review of the sample of recent plans from selected cities, we can draw the following conclusions:

1. City plans have become exceedingly significant instruments for coping with risk in general and the risk stemming from climate change in particular. Nonetheless, the same risks facing the same places have typically been addressed separately by different bodies of knowledge, reflecting the fact that the scholarship has yet to incorporate an understanding that these city plans have the capacity to more effectively address the challenges of bringing different policies and measures under one roof. Plans can bring together mitigation, adaptation, social, economic, and spatial measures and policies into integrated focus under a single plan. For example, mitigation, adaptation, and land use planning may prove synergistic within these plans in reducing transport energy consumption and limiting exposure to floods, or in generating building codes to reduce heating energy consumption and enhance robustness to heat waves. In this way, urban planning provides a context for synergy (IPCC 2012), and the recent city plans may serve as a vehicle for the synergetic effort required to meet the challenge at hand. One of my main conclusions here

is that spatial planning is essential in the efforts of cities to cope with risk and threats. Moreover, city planning and development have an important role to play in contending with the future impact of climate change, the complexities and uncertainties of which pose new theoretical and practical challenges.

2. In the context of planning the risk city, it is clear that different perceptions of risk inform and induce different planning practices in different cities.

3. Some cities have used their plans to articulate their view that a major risk with which they must cope is the risk stemming from climate change and environmental hazards. These cities have seized the opportunity presented by increasing knowledge and awareness of climate change to propose new inclusive plans for their cities. The resulting plans call for coping with climate change, and, at the same, for re-planning and restructuring the city and developing its social and economic spaces. These plans, which have been issued only recently and only by a handful of cities, promote a more inclusive planning practice that takes climate change more seriously, in addition to the further integration of spatial, social, and economic policies.

4. The other cities, which I assume are representative of the vast majority of cities around the world (in Russia, China, and other developing countries), appear to perceive risk differently: that is, as related not to climate change but to future growth opportunities. Their concern is with expansion, economic development, and international competition. "Growth"–the magic word of most of the plans for developed and developing cities–is the inherent mission of the vast majority of the plans, even of those who regard climate change risk as a major concern. As a result, the plans of London, Beijing, Amman, Delhi, Moscow, and many others cast expansion and growth as a major concept of development for their cities.

5. Cities with growth concerns that are not climate change-oriented but related to housing, urbanization, transportation, physical infrastructures, and restructuring continue to apply traditional modern planning approaches (see Figs. 6.1, 6.2, 6.3 and 6.4). By traditional planning approaches I mean planning regarding land use, zoning, urban spatial expansion, transportation system expansion, the development of networks of roads for private vehicles, and the establishment of new industrial areas—all without integrating concepts of sustainability or climate change concerns. The plans of Amman, Moscow, and Beijing seek to enhance growth and economic development without serious consideration of environmental concerns and without the use of sustainable transportation planning, green building, mitigation codes for new buildings, the renovation of existing buildings aimed at reducing the use of energy, and renewable energy. In this way, the recent plans of Amman, Moscow, Delhi, and Beijing resemble plans from the 1920s and the 1950s. Indeed, the vast majority of cities have ignored the sustainable planning approaches and measures that humanity has been busy developing over the last two decades (at least).

6. Cities that take climate change seriously have applied a broad range of mitigation measures aimed at GHG emission reduction. Mitigation appears to be an

easy-to-tackle mission for many of the plans. On the whole, these cities tend to analyze the distribution of GHG emission sources within their cities and then build mitigation policies to target each single source. In terms of eco-form, the plans promote greater compactness and density, enhanced mixed land use, sustainable transportation, greening, and renewal and utilization.

7. Nonetheless, the cities have been neither productive nor creative in the undertaking of adaptation. That is to say, with a number of slight differences between cities, they have all failed in their adaptation approaches. The imperative conclusion is that our cities are not doing all they can to fortify themselves against uncertainties, climate change, and natural and environmental hazards.

8. The plans fail to effectively integrate civil society, communities, and grassroots organizations into the process. The lack of a systematic procedure for public participation throughout cities' neighborhoods and among different social groupings and other stakeholders is a critical shortcoming, particularly during the current age of climate change uncertainty.

9. In order to meet the challenges posed by climate change in the current context of unprecedented uncertainty, planners are in need of a more coordinated, holistic, and multidisciplinary approach. Few cities, however, have thus far invested effort in achieving integrative urban governance. In some countries, another factor resulting in the failure of cities to do so has been the existence of a highly centralized national political system, such as that present in China and Russia.

10. Some may argue that local governments operate under many constraints, resulting in city plans that represent a bland path of least resistance, and that we should therefore not put great faith in the planning documents themselves. I argue that planning should be taken seriously in the context of climate change due to their unique power to integrate policies of mitigation, adaptation, land use, and other related urban measures within one statutory, binding document: the city plan.

11. In many countries, decisions are made at the national level and the influence of cities can be severely limited. Moreover, in many cases, local governments are not the provisioning stakeholders for critical services, as in the case of power utilities, which are not operated by most local governments. Cities need to advance planning horizons and ensure that they are being addressed by service providers and the upper echelons of local government. Local governments only exercise explicit control over the activities for which they are directly responsible, which typically account for only a small percentage of a city's overall GHG emissions.

12. Ultimately, our cities are neither properly nor effectively fulfilling the critical role they should be playing in coping with the risk and uncertainties facing their own residents. For this reason, these cities (and particularly the large ones) may end up being a deathtrap for millions of residents when disasters occur (Fig. 6.5).

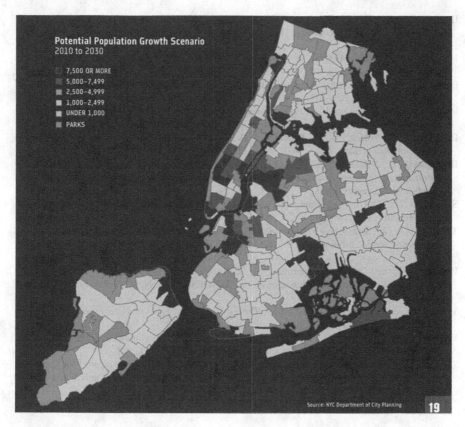

Fig. 6.5 Infill and densification according to New York city PlaNYC 2030

References

ARUP., & C40. (2011). Climate action in megacities: C40 cities baseline and opportunities. Version 1.0 June 2011. file:///C:/Users/use/Downloads/ArupC40ClimateActionInMegacities%20(1).pdf

Barnett, J., & Adger, N. (2005). *Security and climate change: Towards an improved understanding*. Paper Presented at the Human Security and Climate Change Workshop, Oslo, June 21–23, 2005. http://www.gechs.org/downloads/holmen/Barnett_Adger.pdf.

Bicknell, J., Dodman, D., & Satterthwaite, D. (Eds.) (2009). *Adapting Cities to Climate Change: Understanding and addressing the development challenges*. London: Earthscan.

Broto, V. C., & Bulkeley, H. (2013). A survey of urban climate change experiments in 100 cities. *Global Environmental Change, 23*(1), 92–102.

Bulkeley, H. (2013). *Cities and climate change*. London and New York: Routledge.

IPCC (2012). Managing the Risks of Extreme Events and Disasters to Advance Climate Change Adaptation. A Special Report of Working Groups I and II of the Intergovernmental Panel on Climate Change [C.B. Field, V. Barros, T. F. Stocker, D. Qin, D. J. Dokken, K. L. Ebi, M. D. Mastrandrea, K. J. Mach, G.-K. Plattner, S. K. Allen, M. Tignor, and P. M. Midgley (Eds.)]. Cambridge, UK, and New York, NY, USA: Cambridge University Press

IPCC—Intergovernmental Panel on Climate Change (2014) *Climate change 2014: Impacts, adaptation, and vulnerability*. http://ipccwg2.gov/AR5/images/uploads/IPCC_WG2AR5_SPM_Approved.pdf.

Kennedy, C. A., Demoullin, S., & Mohareb, E. (2012). Cities reducing their greenhouse gas emissions. *Energy Policy, 49*, 774–777.

Leichenko, R. (2011). Climate change and urban resilience. *Current Opinion in Environmental Sustainability, 3*(3), 164–168.

Romero-Lankao, P., & Qin, H. (2011). Conceptualizing urban vulnerability to global climate and environmental change. *Current Opinion in Environmental Sustainability, 3*(3), 142–149.

Rosenzweig, C., Solecki, W. D., Hammer, S. A., & Mehrotra, S. (2010). Cities lead the way in climate-change action. *Nature, 467*, 909–911.

Rosenzweig, C., Solecki, W. D., Hammer, S. A., & Mehrotra, S. (2011). *Climate change and cities: First assessment report of the urban climate change research network*. Cambridge, UK: Cambridge University Press.

Solecki, W. (2012). Urban environmental challenges and climate change action in New York City. *Environment and Urbanization, 24*, 557–573.

Tkachenko, L. (2013). *Moscow's master plan 2025*. Moscow: Institute of Moscow City Master Plan. http://www.scribd.com/doc/134535346/Moscows-Master-Plan-2025-by-Ludmila-Tkachenko#scribd.

UNDESA—United Nations Department of Economic and Social Affairs. (2011). *World urbanization prospects, population division*, UNDESA, and New York city. Available at www.un.org/esa/population.

Vale, J. L., & Campanella, T. J. (2005). *The resilient city: How modern cities recover from disaster*. New York: Oxford University Press.

WRI/WBCSD GHG Protocol. (2014). *The global protocol for community-scale greenhouse gas emission inventories*. http://ghgprotocol.org/files/ghgp/GHGP_GPC.pdf.

Zhao, J. (2011). *Climate change mitigation in Beijing, China*. Case Study Prepared for Cities and Climate Change: Global Report on Human Settlements 2011. http://www.unhabitat.org/grhs/2011.

Chapter 7
The Risk City Resilience Trajectory

7.1 Introduction

In light of the complex challenges facing our cities and our concern for the future of their human communities, this chapter poses some critical questions about the resilience of contemporary cities: How resilient are they? That is, are they ready to face multiplicity of environmental, economic, social, and security challenges and uncertainties of the future? How resilient will they be in the future? Or, to phrase it differently, what is their resilience trajectory? In the international context, do the resilience trajectories and settings of Western cities differ from those of other world cities? What kind of resilience should we aspire to in our ideal city? How can we compare the resilience levels of different cities? Perhaps most importantly, in light of the multiplicity of challenges currently facing our cities, how should we go about understanding and analyzing urban resilience at the present and in the future? Moreover, since human action contributes to the altering of the ecosystem locally and globally (Folke et al. 2011; Chapin et al. 2011), how resilient should cities be in order to contribute to environmental protection and sustainability?

This chapter aims theoretically to contribute to this critical planning field by investigating the phenomenon of city resilience and developing a more systematic, multidisciplinary conceptual framework for understanding the complexity of urban resilience and exploring its trajectories.

A review of the literature reveals a marked absence of theory on urban resilience. Only in recent years have scholars begun to investigate and write about urban resilience. Up until quite recently, most of the literature on resilience to environmental threats has focused on disaster areas and disaster stricken communities and on poor rural communities in developing countries. Thus, there is as yet no comprehensive conceptual framework of urban resilience that takes into account not only environmental risks, but also social, economic, and security risks and challenges. Nor has any method of assessment, which encompasses all these threats been developed. We thus have no means of systematically comparing the levels and kinds of resilience of different cities.

7.2 The Problem of Resilience

Although a literature review reveals an important emerging scholarship on urban resilience, most studies on the subject make use of general, vague, and confusing terminology. They fail to conceptualize and theorize the phenomenon in a systematic manner. Therefore, this chapter aims to fill the theoretical and practical gaps and answer the critical question regarding what cities and their urban communities should do in order to move towards a more resilient future state.

The concept of resilience, in the urban context, was borrowed from studies on the manner in which ecological systems cope with stresses and disturbances caused by external factors (Davic and Welsh 2004). From an ecological perspective, Holling (1973), who may be the first to define it (Barnett 2001; Carpenter et al. 2001), suggests that resilience is "the persistence of relationships within a system" and "the ability of these systems to absorb changes of state variables, driving variables, and parameters, and still persist" (Holling 1973: 17). In other words, resilience is "the capacity of a system to undergo disturbance and maintain its functions and controls" (Gunderson and Holling 2001).

Recently, the concept has also been applied to human social systems (Adger 2000; Pelling 2003; Leichenko 2011); ecological urban resilience (Andersson 2006; Barnett 2001; Folke 2006; Ernstson et al. 2010; Maru 2010); economic recovery (Rose 2004; Martin and Sunley 2007; Pendall et al. 2010; Pike et al. 2010; Simmie and Martin 2010), disaster recovery (Colten et al. 2008; Cutter et al. 2008; Pais and Elliot 2008; Vale and Campanella 2005; Coaffee and Roger 2008; UNISDR 2010), and urban security and resilience against post-September 11th terrorism (Coaffee 2006, 2009). Inspired by the concept of the resilient ecosystem, "resilience means the ability of a system, community or society exposed to hazards to resist, absorb, accommodate to and recover from the effects of a hazard in a timely and efficient manner, including through the preservation and restoration of its essential basic structures and functions" (UNISDR 2010: 13).

Apparently, a striking weakness of the scholarship on the subject is its lack of multifaceted theorizing and the fact that it typically overlooks the multidisciplinary and complex nature of urban resilience. Because city resilience is a complex, multidisciplinary phenomenon, focusing on a single or small number of contributing factors ultimately results in partial or inaccurate conclusions and misrepresentation of the multiple causes of the phenomenon. Folke et al. (2010) suggest that resilience is about dynamic and complex systems, which is characterized by multiple pathways of development, interacting periods of gradual and rapid change, feedbacks and non-linear dynamics, thresholds, tipping points and shifts between pathways, and how such dynamics interact across temporal and spatial scales

(Folke et al. 2011: 721). Godschalk (2003: 14) contends that if we want to take urban resilience seriously, we need to build the goal of a resilient city in a multi-disciplinary manner. Little (2004) posits that resilience is about more than just physical robustness and will be less effective if restricted to a narrow discipline. Moreover, some scholars argue that critical urban issues "are typically treated as independent issues," and that "this frequently results in ineffective policy and often leads to unfortunate and sometimes disastrous unintended consequences" (Bettencourt and Geoffrey 2010). In this context, Bettencourt and Geoffrey (2010: 912) conclude, "Developing a predictive framework applicable to cities around the world is a daunting task, given their extraordinary complexity and diversity." Leichenko (2011: 164) concludes that urban resilience studies are grounded in a diverse array of literatures, and "while there is much overlap and cross-fertilization among these different sets of literature, each emphasizes different facets of urban resilience and each focuses on different components of cities and urban systems."

Another gap in the literature is related to measuring resilience and how to assess a system's resilience in general and urban resilience in particular. Mostly, the literature of resilience measurements has focused on ecosystems, and suggests quantitative indicators for such assessment. According to Gunderson and Holling (2001), resilience is measured by the magnitude of disturbance that can be experienced without the system flipping into another state and within which the system can absorb and still persist. Carpenter et al. (2001) suggest measurement of socio ecological systems (SES) that focuses on its capacity. It appears that the resilience concept has been applied mostly to understand social–ecological systems and dynamics in areas that suffer disaster, rural communities in developing countries, and for improving livelihoods (Chapin et al. 2009; Eakin and Wehbe 2009; Enfors and Gordon 2008; Folke et al. 2011; McSweeney and Coomes 2011; Walker et al. 2006; WRI/WBCSD GHG Protocol 2014). In conclusion, the literature on measuring resilience overlooks cities and ordinary communities (see also Castello 2011).

One example of this type of treatment is the new campaign launched by the United Nations International Strategy for Disaster Reduction in 2010, entitled *Making Cities Resilient* (UNISDR 2010). The campaign aims to "promote awareness and commitment for sustainable development practices that will reduce disaster risk and increase the wellbeing and safety of citizens—'to invest today for a better tomorrow'" (UNISDR 2010). The UNISDR proposes a general and limited scope checklist of ten essentials to empower local governments and other agencies to implement the *Hyogo Framework for Action 2005–2015*. This framework focuses on "Building the Resilience of Nations and Communities to Disasters" (UN/ISDR 2005), which was adopted by 168 governments in 2005.

In *Resilient Cities*, Newman et al. (2009) also focus on only one dimension of resilience: the oil crisis. In this context, they point out that "a danger that few think about with such immediacy is the threat of the collapse of our metropolitan regions in the face of resource depletion—namely, the reduction in the availability of oil and necessary reduction in all fossil fuel use to reduce human impact on climate change" (2009: 2). In this way, their book focuses less on urban resilience and more on "the challenges posed to metropolitan areas in the face of responding to their increased carbon footprint, dependence on fossil fuels, and impact on the irreplaceable natural resources" (2009: 2).

In *The Resilient City*, Vale and Campanella (2005) focus on the narratives of resilience, the symbolic dimensions of disaster and recovery, and the politics of reconstruction. They argue that, to understand urban resilience is to understand the ways in which human narratives are constructed to interpret the meanings of urban reconstruction. *The Resilient City* by Walisser et al. (2005), which was prepared by the Vancouver Working Group for the 2006 World Urban Forum, explores the resilience of small Canadian communities dependent on single resource industries by examining how they have coped with the economic and social pressures arising from widespread closures.

In summary, the major theoretical challenge regarding urban resilience facing many scholars today appears to be the development of a multidisciplinary theory that integrates a variety of urban dimensions such as social, economic, cultural, environmental, spatial and physical infrastructure, into a unified conceptual framework for understanding the resiliency of cities and how they should move towards a more resilient state. Therefore, this paper aims to fill the theoretical and knowledge-based gaps in this critical field by investigating the phenomenon of city resilience and developing a new multidisciplinary conceptual framework for understanding the complexity of urban resilience. In other words, this paper seeks to construct a more rigorous, careful basis for promoting and assessing resilience of cities.

7.3 Conceptual Framework for the Resilient City

I suggest that a city's resilience is related to social, economic, environmental, and security resilience. Urban resilience is the totality of its resilience to the environmental, economic, social, and security threats that cities face. Environmental threats consist of natural hazards and disasters, as well as the impacts of climate change (e.g., higher temperatures and sea levels; increased strains on materials and equipment, higher peak electricity loads, transport disruptions, and increased need for emergency management). Economic threats refer to such things as collapsed industries, dramatic job loss, and housing crises, stemming from a range of causes. Social threats include such things as socio-economic disparities, socio-spatial segregation, and sectarian conflicts. Security threats encompass violence and terrorism. Urban resilience thus has four dimensions (see Fig. 7.1): (a) *environmental resilience*; (b) *social resilience*; (c) *economic resilience*; and (d) *security resilience*. Each dimension refers to the city's ability to prepare for, cope with, and recover from the threat at issue. In this book, I focus on city resilience that is related to environmental crisis and climate change impacts and threats.

In my theorization of the *Risk City Resilience Trajectories,* or the Trajectories of City Resilience, I adopt the ontological conceptualization of *planes of immanence* and the philosophy behind the term *concept* used by Deleuze and Guattari. My aim

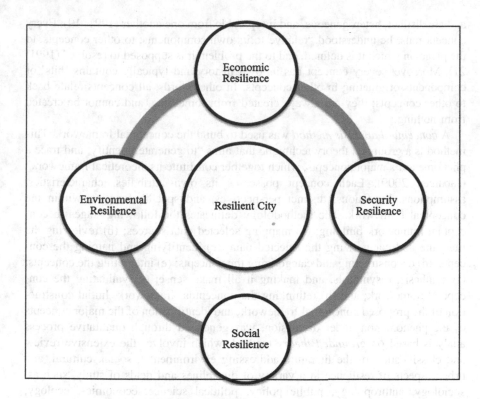

Fig. 7.1 Dimensions of urban resilience

is to build a conceptual framework for *Risk City Resilience Trajectories* or *Trajectories of City Resilience*. The conceptual framework for *RCRT* is a plane of immanence that is "an object of construction" (Bonta and Protevi 2004: 62–63), defined not by what it "contains but by the forces that intersect it and the things it can do" (Kaufman 1998: 6). In philosophical terminology, immanence refers to "the act of being within a conceptual space" (Bonta and Protevi 2004, 98). In this manner, the conceptual framework of *Trajectories of City Resilience* refers to "a network, or 'a plane,' of interlinked *concepts* that together provides a comprehensive understanding of a phenomenon" (see Jabareen 2009: 51). This plane is composed of concepts that "can be abstracted from bodies and states of affairs" (Bonta and Protevi 2004: 31). A conceptual framework, however, is not merely a collection of concepts but rather a construct composed of 'consistent' concepts in which each plays an integral role and is intrinsically linked to the others. This enables it to better provide "not a causal/analytical setting but, rather, an *interpretative approach* to social reality" and to our understanding of the multiple and interlinked concepts it encompasses (Jabareen 2009: 51).

The Definition of 'Concept': According to Deleuze and Guattari (1991: 15), "every concept has components and is defined by them" and "there is no concept with only one component." These components define the consistency of the concept and are distinct, heterogeneous, and inseparable from one another (1991: 19). Every concept must be understood "relative to its own components, to other concepts, to the plane on which it is defined, and to the problem it is supposed to resolve" (1991: 21). Moreover, every concept has its own history and typically contains 'bits' or components originating in other concepts. In other words, all concepts relate back to other concepts; they are always created from something, and cannot be created from nothing.

A *conceptual analysis method* was used to build the conceptual framework. This method is a grounded theory technique that aims "to generate, identify, and trace a phenomenon's major concepts, which together constitute its theoretical framework" (Jabareen 2009). Each concept possesses its own attributes, characteristics, assumptions, limitations, distinct perspectives, and specific function within the conceptual framework. The methodology delineates the following stages in conceptual framework building: (a) mapping selected data sources; (b) reviewing the literature and categorizing the selected data; (c) identifying and naming the concepts; (d) deconstructing and categorizing the concepts; (e) integrating the concepts; (f) synthesis, resynthesis, and making it all make sense; (g) validating the conceptual framework; and (h) rethinking the conceptual framework. Initial construction of the proposed conceptual framework, and identification of the major concepts of the phenomenon under discussion, was generated through qualitative process analysis based on *grounded theory* method, which involves the extensive review and classification of the literature addressing environmental, social, cultural, and urban aspects of resilience in a variety of disciplines and fields of study, such as: sociology, anthropology, public policy, political science, economics, ecology, geography, and urban planning. This broad multidisciplinary framework is intended to ensure that the theory generated is relevant to as many disciplines as possible, thus enabling them to expand the respective theoretical perspectives with which they approach the phenomenon.

As Fig. 7.1 demonstrates, the analysis reveals a conceptual framework that is composed of four main interrelated concepts and their components.

7.3.1 Concept 1: Vulnerability Analysis Matrix

This concept is critical and significant for the resilient city and for its contribution to the spatial and socio-economic mapping of future risks and vulnerabilities. The role of the Vulnerability Analysis Matrix is to analyze and identify types, demography, intensity, scope, and spatial distribution of environmental risk, natural disasters, and future uncertainties in cities. In addition, this concept seeks to address how hazards, risks, and uncertainties affect various urban communities and urban groups. In the context of climate change, vulnerability refers to the "degree to which a system is susceptible to, and unable to cope with, adverse effects of climate change, including climate variability and extremes. Vulnerability is a function of a system's exposure, its sensitivity, and its adaptive capacity" (CCC 2010: 61).

The concept of the Vulnerability Analysis Matrix is composed of four main components that determine its scope, environmental, social, and spatial nature. These four components are:

(a) **Demography of Vulnerability**: This component assesses and examines the demographic and socio-economic aspects of urban vulnerability. It assumes that there are individuals and groups within all societies who are more vulnerable than others and lack the capacity to adapt to climate change (Schneider et al. 2007: 719). Demographic, health, and socio-economic variables affect the ability of individuals and urban communities to face and cope with environmental risk and future uncertainties. Many variables affect the vulnerability of individuals and communities: income, education and language skills, gender, age, physical and mental capacity, accessibility to resources and political power, and social capital (Cutter et al. 2003; Morrow 1999; Ojerio et al. 2010; United Nations Division for the Advancement of Women 2001). As a result, socio-economically weak communities are more vulnerable to suffer negative impacts, including property loss, physical harm, and psychological distress (Ojerio et al. 2010; Fothergill and Peek 2004).

(b) **Informality**: This concept assesses the scale and social, economic, and environmental conditions of informal urban spaces. Informal spaces are unplanned, chaotic, and disorderly (Roy 2010) and it is assumed that the scale and human condition of informal places within a city have a significant impact on its vulnerability.

(c) **Uncertainty**: This component has a critical impact on urban vulnerability and requires the assessment of environmental risks and hazards that are difficult to predict but must be taken into account in city planning and risk management.

(d) **Spatial Distribution of Vulnerability**: This component assesses the spatial distribution of risks, uncertainties, vulnerability and vulnerable communities in cities. Environmental risks and hazards are not always evenly distributed geographically, and some communities may be affected more than others. For example, those who are close to the shore may be affected more harshly by tsunamis than others. Mapping the spatial distribution of risks and hazards is critical for planning and management at the present and for the future. In addition, the communities that are most vulnerable to climate change impacts are usually those who live within more vulnerable, high-risk locations that may lack skills, adequate infrastructure and services (Satterthwaite 2008).

7.3.2 Concept 2: Urban Governance

This concept contributes to the governance of urban resilience. It focuses on the governance culture, processes, arena and roles of the resilient city. It is hypothesized that a more resilient city is one with inclusive decision making processes in the realm of planning, open dialogue, accountability, and collaboration. It is one in which people and local stakeholders, including the private sector, various social groups, communities, civil society and grassroots organizations participate. A more resilient city is one in which governance is able to quickly restore basic services and resume social, institutional and economic activity after a disastrous events. Weak governance, on the other hand, lacks the capacity and competence to engage in participatory planning and decision making, and will typically fail to meet the challenges of resilience as well as increase the vulnerability of much of the urban population (Albrechts 2004; Healey 2007, 2010; UNISDR 2010; Dodman et al. 2009). This concept suggests that in order to cope with uncertainties, risks and hazards that cities and communities may face, there needs to be a shift in urban governance. This shift will make urban governance more integrative, deliberative, and socially and eco-economically sound. Accordingly, this concept is composed of the three following components:

(a) *Integrative Approach*: In order to enhance the urban governance of climate change and cope with environmental disasters and uncertainties, there is a need to expand and improve local capacity through increasing knowledge, providing resources, establishing new institutions, enhancing good governance, and granting more local autonomy (Allman et al. 2004; Bai 2007; Corfee-Morlot 2009; Harriet 2010; Holgate 2007; Lankao 2007; Bulkeley et al. 2009; Kern 2008: 56). This component represents the integrative framework for city planning and adaptive management under conditions of uncertainty and the spectrum of collaboration that a plan proposes.

(b) *Equity*: This component encompasses social issues such as poverty, inequality, environmental justice, and public participation in decision-making and space production. Therefore, it plays a central role in shaping a city's resilience.

(c) *Ecological Economics*: This component contributes to the assessment of the economic aspects of urban resilience and the economic engines cities put in place to meet climate change objectives and environmental hazard mitigation and reduction. It suggests that only environmentally sound economics can play a decisive role in achieving urban resilience and climate change objectives in a capitalist world. The idea is to create opportunities to integrate resilience planning, protection and development approaches into the city economic development decisions and strategies; and to shape reforms in the area of investment, insurance and risk management related to natural disasters and other emergencies (NYS 2100 2013: 10).

7.3.3 Concept 3: Prevention

This concept suggests that in order to move towards greater urban resiliency and less vulnerability, cities need to seek to prevent environmental hazards and climate change impacts. This is composed of three main components that aim at preventing future catastrophes. These components assess urban mitigation policies to reduce hazards, involve the spatial restructuring of the city in order to prepare it for a future environmental disaster, and seek alternative clean energy. These components are:

(a) *Mitigation*: this component assesses policies and actions that aim to reduce greenhouse gas emissions (GHG).
(b) *Restructuring*: This concept represents the ability and flexibility of a city to restructure itself in order to face social, environmental, and economic challenges. For example, the shift towards a knowledge-based economy and the emphasis on the production, trade, and diffusion of knowledge has triggered specific spatial structural transformation in cities (Cooke and Piccaluga 2006). At the same time, cities vary in their strategies and ability to restructure, as well as their readiness to address uncertainties.
(c) *Applying Alternative Energy*: It is hypothesized, that cleaner, more efficient, and renewable use of energy is a key to achieving greater city resilience (CCC —Committee on Climate Change 2010). This concept suggests that energy should be based on new low-carbon technologies in order to meet emissions reduction targets.

7.3.4 Concept 4: Uncertainty-Oriented Planning

This concept suggests that planning should be uncertainty-oriented rather than adapting the conventional planning approaches. It suggests that climate change and its resulting uncertainties challenge the concepts, procedures, and scope of conventional approaches to planning, creating a need to rethink and revise current planning methods. Alternatively, this paper acknowledges not only the physical dimensions of planning to prevent hazards, but suggests that planning has a wider role to play that is closely associated with uncertainties. Abbott identified this role well when he said, "planning means, essentially, controlling uncertainty—either by taking action now to secure the future, or by preparing actions to be taken in case an event occurs" (2005: 237). Moreover, "the uncertainties about the overall impacts of climate change and accelerated sea-level rise are magnified many times over due to incomplete knowledge of individual ecosystems, patterns of causality and interaction between ecosystems, and patterns of causality and interaction between social and ecological systems" (Barnett 2001: 2–3).

The uncertainty concept is composed of three interrelated components as follows:

(a) *Adaptation*: In order to counter climate change, there is a crucial need for uncertainty management that includes adaptation policies. Obviously, a resilience approach to a climate change adaptation should address uncertainties and limit impacts even if magnitude and direction are uncertain or unknown (Wardekker et al. 2010: 995; Dessai and van der Sluijs 2007). Adaptation is the task of modifying ecological and social systems to accommodate climate change impacts, such as accelerated sea-level rise, so that these systems can persist over time (Barnett 2001). Barnett argues that adaptation is hard to grasp because it demands system-wide analysis and intervention.

The new urban uncertainties posed by climate change challenges the concepts, procedures, and scope of planning. In order to cope with the new challenges, planners must develop a greater awareness and place mitigation and policies for adaptation, or actual adjustments that might eventually enhance resilience and reduce vulnerability to expected climate change impacts, as the focus of the planning process (Adger et al. 2007: 720). When we adopt adaptation measures, we acknowledge that the climate will continue to change and that we must take measures to reduce the risks brought about by these changes (Priemus and Rietveld 2009). From this perspective, adaptation to climate change must be considered as indispensable (Vellinga et al. 2009). Moreover, adaptation planning, practices and policies should also consider statistical uncertainties, scenario uncertainties, or sometimes recognize ignorance (Walker et al. 2003). Wardekker et al. (2010) suggest that "scenario uncertainty" stems from limited predictability of the future, and that resilient systems can cope with statistical uncertainty. In addition, resilient systems should be able to deal with a continuous range of conditions, rather than only the 'average.'

This concept evaluates a plan's adaptation strategies (*ex-post* and *ex-ante*) and policies in addition to the planning strategies for addressing future uncertainties stemming from climate change. In plan evaluation, the following questions need to be answered in order to fit within this concept: Does the plan include development projects for infrastructure design in order to reduce vulnerabilities and make the city more resilient? Does the plan enhance the city's adaptive planning capacity, or the ability of the planning system to respond successfully to climate variability and change?

(b) *Spatial Planning*: This component assesses the role of planning in transforming the city into a more resilient state. It "is the provision… of future 'certainty' in a complex, unstable, dynamic and inherently uncertain world" (Gunder and Hillier 2009: 23). Therefore, planning plays a more central role in shaping all dimensions of the built environment, including in physical security and environmental and socio-spatial policies, and has a major impact on city resilience.

In order to reduce community vulnerability to natural hazards, planning scholars have advocated using land use management, and building and site design codes to regulate development in hazard prone areas (Burby et al. 2000; Godschalk et al. 1999; Nelson and French 2002; Zhang 2010). However, planning should expand its scope to include prediction and anticipation of risks and uncertainties as well as provide ways to cope with them.

(c) *Sustainable Urban Form*: The physical form of a city affects its habitats and ecosystems, the everyday activities and spatial practices of its inhabitants and, eventually, climate change (Jabareen 2006). This component assesses spatial

planning, architecture, design, and the ecologically desired form of the city and
its components. I (Jabareen 2006) suggest the following set of nine planning
typologies, or criteria of evaluation, which are helpful in evaluating plans from
the perspective of eco-form:

1. *Compactness*: This refers to urban contiguity and connectivity.
2. *Sustainable Transport*: This suggests that planning should promote sustainable modes of transportation through traffic reduction, trip reduction, the encouragement of non-motorized travel, transit-oriented development, safety, equitable access to all, and renewable energy sources.
3. *Density*: High density planning can save significant amounts of energy, yet, it maybe problematic to safety.
4. *Mixed Land Uses*: This indicates the diversity of functional land uses, such as residential, commercial, industrial, institutional, and transportation.
5. *Diversity*: This is a multidimensional phenomenon that promotes other desirable urban features, including a larger variety of housing types, building densities, household sizes, ages, cultures, and incomes (Turner and Murray 2001: 320).
6. *Passive Solar Design*: This aims to reduce energy demands and provide the best use of passive energy through specific planning and design measures,

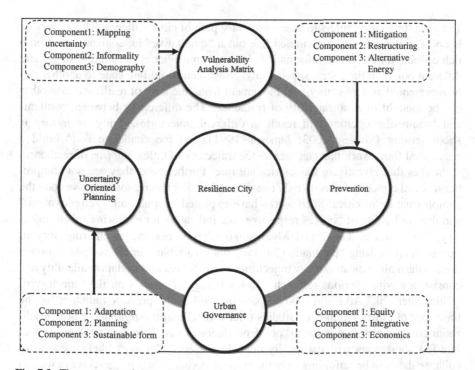

Fig. 7.2 The concepts of the resilient city

      such as orientation, layout, landscaping, building design, urban materials,
      surface finish, vegetation, and bodies of water.
7. *Greening*: This makes positive contributions to many aspects of the urban
      environment, including biodiversity, the lived-in urban environment, urban
      climate, economic attractiveness, community pride, health and education.
8. *Renewal and Utilization*: These refer to the process of reclaiming the many
      sites that are no longer appropriate for their originally intended use and can
      be used for a new purpose, such as brownfields (Fig. 7.2).

7.4 Risk City Resilience Trajectories

I employ the term *risk city trajectory* in conjunction with the term resilience to yield
the concept of *Risk City Resilience Trajectories* (RCRT), which can be used to
assess and explain the direction, patterns, and properties of a given city with regard
to its past, current, and future resilience setting.

7.4.1 City Trajectory

The term 'trajectory' refers to an evolutionary path of city resilience. This trajectory
is not a mathematically determined line but a "qualitative" assessment based on a
rich collection of data and information about urban resiliency at different stages and
different points in time for a specific city. Movement from one state to another can
be represented as a trajectory, and movement from one state of resilience to another
can be thought of as a trajectory of resilience. The differences between 'resilient'
and 'non-resilient' cities will result in different trajectories. Truly, *trajectory* is
about sensing (Massey 2003; Jameson 1991), and the challenge is to build a
conceptual framework that can 'sense' the trajectory of cities in terms of resilience.
    Studies that investigate trajectories are rare. Furthermore, they are not compre-
hensive and typically focus on only one dimension. For this reason they overlook the
complex nature of cities. Such works have explored "population," "employment,"
and the availability of "oil" as respective sole indicators for exploring urban trajectory. For example, Turok and Mykhnenko (2007) seek the recent trajectory of
European cities using "population" as their main variable. For them, "population is
used as the main indicator of city trajectories partly for reasons of data availability and
consistency with previous research." In his trilogy of articles on the trajectory of
"Cities after Oil," Atkinson (2007, 2008) uses oil as the primary indicator for the
trajectory of cities. Atkinson's studies offer a detailed investigation of how civili-
zations and cities collapse due to "our dependence of vast throughputs of energy that
in a few short years will start to dry up." He presents "the most likely scenario of
collapse that will be unfolding over the coming decades" and suggests ways we can

survive the consequences (Atkinson 2007). He concludes by envisioning the stages of collapse of 'modern' civilization over the coming decades (Atkinson 2008).

One striking weakness of the existing scholarship on the *city resilience* is its lack of multifaceted theorizing. Another is the fact it tends to overlook the multidisciplinary and complex nature of urban resilience and city trajectories. Because city trajectories and urban resilience are complex multidisciplinary phenomena, focusing on only one or a small number of contributing factors ultimately leads to partial or inaccurate conclusions. Moreover, because we are dealing with a multidisciplinary dynamic phenomenon, and because we are concerned with "relationships and process of unpredictable movement or emergence" (Hillier 2010: 499), complexity thinking and a complexity approach are highly relevant to this study. Still, "there is no all-embracing complexity theory" (Hillier 2010: 499; see Urry 2005) in general, and no available methodology for approaching the multiplicity of urban resilience trajectories in particular. Moreover, complexity theory demands a broad and open-minded approach to epistemological positions and methodological strategies. As Richardson and Cilliers (2001: 12) argue: "If we allow different methods, we should allow them without granting a higher status to some of them. Thus, we need both mathematical equations and narrative descriptions. Perhaps one is more appropriate than the other under certain circumstances, but one should not be seen as more scientific than the other (Richardson and Cilliers 2001: 12)." In other words, complexity implies methodological pluralism (Richardson 2005).

Assessment of Concept Attributes: In order to assess the role of each of the framework's component concepts in shaping city resilience, the study will develop indicators and measures according to the following assumptions and procedures:

- Every component incorporates its own sub-concepts. These component concepts are distinct, heterogeneous, and inseparable from one another (Deleuze and Guattari 1991: 15–19). As an example, we take two components of the concept of *spatial planning* employed in this proposal: (a) *scope of planning*, and (b) *nature of planning*.
- Each component can be measured on a scale. In the case of the concept of *spatial planning*, *scope of planning* can be understood on a scale ranging between '*opportunistic planning*,' which lacks a coherent vision of the city, and '*masterminding planning*,' which crafts a comprehensive trajectory for the city (see Boyle and Rogerson 2001: 405).
- A component may be measured both qualitatively and quantitatively, depending on its definition and the availability of data.
- Overall, a specific concept's contribution to the resilience of a city is the sum of the contribution of its components.
- For the sake of uniformity, scales and measurements will be normalized and standardized.
- Each concept has a past, a present, and a future trajectory, as do each of its components. In this proposal, each component will be devised and measured according to a scale ranging from a very low contribution to resilience to a very high contribution to resilience.

The **Risk City Resilience Trajectory** (RCRT) is a complex phenomenon: non-linear, fundamentally non-deterministic, dynamic in structure, and uncertain in nature. It is affected by a multiplicity of economic, social, spatial, and physical factors; its planning involves a wide range of stakeholders, including civil society, local and national governments, the private sector, and various professional communities; and it affects a variety of urban communities and city residents. By nature, working on urban resilience requires "complex thinking and complex methods" (de Roo and Juotsiniemi 2010: 90), and the complexity approach offers a suitable method of generating the kind of insights we seek with regard to the future trajectories of cities. It also forces us to adopt a more holistic view (Batty 2007) (Figs. 7.3 and 7.4).

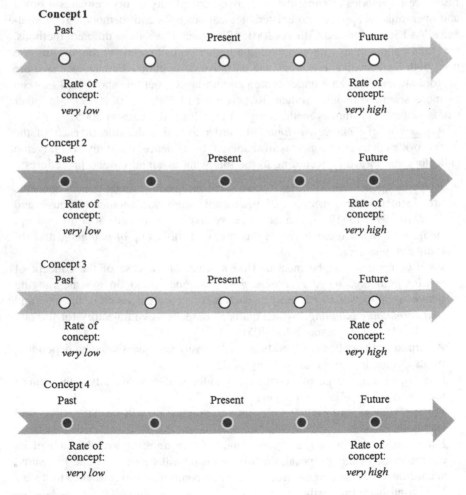

Fig. 7.3 Concepts and their trajectories

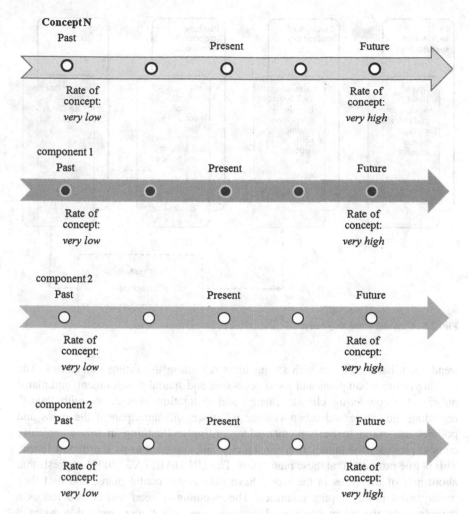

Fig. 7.4 Components of concepts and their trajectories

This framework allows us to figure out how many urban resilience trajectories each city has. In addition, it allows us to understand the strengths and weaknesses for each specific city.

In order to track the city trajectories, we assume that each city has its socio-spatial, economic, political and cultural specifications and contexts. These specifications become the socio-spatial departure "point" of the trajectories. Apparently, each city has its own specifications and therefore has its own "historic" departure point. Then the conceptual framework and its concepts, including the concepts' components, provide us with the existing socio-spatial, environmental, economic, governance, and security settings regarding the city resilience strengths and weaknesses at the present. The conceptual framework provides us the past-present

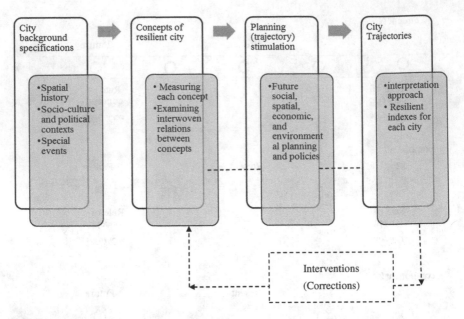

Fig. 7.5 City resilience trajectory

trends of urban resilience with strong hints regarding the future trajectories. The existing future urban plans and socio-economic and spatial development, and plans/ policies for countering climate change and mitigation provide us with "hints" regarding the future and urban visions: what we will anticipate if the plans and policies are going to be implemented accordingly. Planning in this complexity context provides the stimulation for the city and suggests to us a different future. This is true provided that these plans exist. The UN-HABITAT (2009) suggests that about half of the cities in the world have no environmental plans, and that they mostly exist in developing countries. The planning is seen and considered as a stimulator on the urban complex. It is also seen as a future simulation when it suggests future interventions that aim to change the course of business as usual and to change the undesired trajectory of the city.

Apparently, the city trajectory, which represents the behavior of the city (which is a chaotic system), is unpredictable in quantitative detail but its trajectory could be predictable 'qualitatively' (see Protevi 2006) as this chapter assumes (Fig. 7.5).

7.5 Conclusions

The idea behind the *Risk City Resilience Trajectory* is that our cities must learn from the past and the present in order to plan the uncertain futures, since "resilience requires frequent testing and evaluation" (NYS 2100 2013: 7). The learning should be based mainly on our experience and emerging knowledge on vulnerability and adaptation measures. The *Risk City Resilience Trajectory* is to acknowledge the current as well as the future vulnerabilities and risks and to plan the future differently.

Following Superstorm Sandy, Judith Robin, the President of Rockefeller Foundation and Felix Rohatyn state (NYS 2100 Commission 2013: 7) conclude, "We have to build back better and smarter." Convincingly, they contend (Robin and Rohatyn 2013: 10):

> The next century will be defined by the extent to which our communities are resilient to the direct and indirect impacts of a rapidly changing climate and other long-term accelerators of change. We will never be able to predict or prevent all extreme events. But we must not waste the lessons learned and opportunities afforded by these recent storms to chart a course for the State that truly prepares our communities for future eventualities. Planning for a more resilient tomorrow enables the State and its residents to take cost-effective actions and to make investments that will benefit our communities today and far into the future.

The resilient city conceptual framework addresses a critical question of what cities and their urban communities should do in order to move towards a more resilient state in the future. Accordingly, the *Resilient City Framework* is defined as a network, or a theoretical plane of interlinked concepts, that provides a comprehensive understanding of city resilience. The *Resilient City Framework* is composed of four concepts. As described in this paper, each concept consists of specific components that define its nature and assess its contribution to the framework. The contribution of each concept to the framework of urban resilience is the sum of the contributions of its measurable components. Even though various measurement techniques may already exist for some of these components, a systematic approach to the measurement of all components should be prioritized in future research.

Each one of the four *Resilient City Framework* concepts has specific roles and domain in the RCT framework as seen in Fig. 7.1 and Table 7.1. The 'Urban vulnerability matrix analysis' focuses on the governance culture, processes, arena and roles of the resilient city. This concept is critical and significant for the resilient city for its contribution to the spatial and socio-economic mapping of future risks and vulnerabilities. The 'Urban governance' concept contributes to the holistic management of urban resilience. It focuses on urban policies and assumes that there is a significant need for a new approach to urban governance in order to cope with uncertainties and future environmental and climate change impact challenges. Urban governance suggests that the integrative governance approach, deliberative and communicative decision making measures, and ecological economics have a great impact on moving our cities towards improved urban resiliency. Moreover, "building back better demands a focus on increased resilience: the ability of individuals, organizations, systems, and communities to bounce back more strongly from stresses and shocks" (NYS 2100 Commission 2013: 7).

The concept of 'prevention' represents the various components that should be considered in order to contribute to the prevention of environmental hazards and climate change impacts. These components include mitigation measures, adaptation

Table 7.1 Resilient city framework

Concepts	Components	Key questions (measurements)
Concept 1: Urban vulnerability matrix analysis	C1: Uncertainties	C1. What are the hazard and environmental uncertainties?
	C2: Informality	C2. What is the scope, geography, socio-economic, demographic, and physical characters of existing informal settlements in or closed to the city?
	C3: Demography	C3. What is the nature of vulnerable demography in the city by age, gender, health, and other social group?
	C4: Spatiality	C4. What is the spatial distribution of environmental hazards and risks?
Concept 2: Uncertainty oriented planning	C1: Adaptation	C1: What adaptation measures are taken to reduce risks and cope with future uncertainties?
	C2: Planning	C2: How do planning methods cope with uncertainties?
	C3: Sustainable Form	C3: What are characteristics of the existing and planned urban form typologies?
Concept 3: Urban governance	C1: Equity	C1: Who participates in decision-making and planning regarding environmental and uncertainty issues?
	C2: Integrative	C2: Is the urban governance approach integrating institutional, legal, social, economic, and environmental aspects?
	C3: Eco-economics	C3: What is the nature of the existing and planned ecological economy?
Concept 4: Prevention	C1: Mitigation	C1: What mitigation measures are taken to reduce risks and to prepare the city for future environmental hazards?
	C2: Restructuring	C2: What are the proposed or planned spatial, physical, and economic restructuring policies that aim to face the environmental hazards and uncertainties?
	C3: Alternative energy	C3: How does the city address the energy sector and does it propose strategies to reduce energy consumption and to use new alternative and cleaner energy sources?

of clean energy, and urban restructuring methods. The fourth concept, 'uncertainty oriented planning', demonstrates that planning should adapt its methods in order to help cities cope with uncertainties in the future.

Resilient City Framework, the framework for City Resilience and Community Resilience, is a complex phenomenon, non-deterministic, dynamic in structure, and uncertain in nature. It is affected by a multiplicity of economic, social, spatial, and physical factors and its planning involves a wide range of stakeholders. It is important to mention that the proposed *RCPF* is not a deterministic framework, but a dynamic and flexible one that could be modified while respecting its fundamental concepts.

According to the *Resilient City Framework,* a resilient city is defined by the overall abilities of its governance, physical, economic and social systems and entities that are exposed to hazards to learn, be ready in advance, plan for uncertainties, resist, absorb, accommodate to and recover from the effects of a hazard in a timely and efficient manner, including through the preservation and restoration of its essential basic structures and functions.

References

Adger, W. N. (2000). Social and ecological resilience: Are they related? *Progress in Human Geography, 24*(3), 347–364.

Adger, W. N., Agrawala, S., Mirza, M. M. Q., Conde, C., O'Brien, K., Pulhin, J., et al. (2007). Assessment of adaptation practices, options, constraints and capacity. In Parry, M. L., Canziani, O. F., Palutikof, J. P., van der Linden, P. J., & Hanson, C.E. (Eds.), *Climate change 2007: Impacts, adaptation and vulnerability. Contribution of working group II to the fourth assessment report of the intergovernmental panel on climate change* (pp. 717–743). Cambridge, UK: Cambridge University Press.

Albrechts, L. (2004). Strategic (spatial) planning reexamined. *Environment and Planning B: Planning and Design, 31,* 743–758.

Allman, L., Fleming, P., & Wallace, A. (2004). The progress of english and Welsh local authorities in addressing climate change. *Local Environment, 9,* 271–283.

Andersson, E. (2006). Urban landscapes and sustainable cities. *Ecology and Society, 11,* 34.

Atkinson, A. (2007). Cities after oil—2: Background to the collapse of 'modern' civilization. *City, 11*(3), 293–312.

Atkinson, A. (2008). Cities after oil—3 collapse and the fate of cities. *City, 12*(1), 79–106.

Bai, X. (2007). Integrating global environmental concerns into urban management: The scale and readiness arguments. *Journal of Industrial Ecology, 11,* 15–29.

Barnett, J. (2001). Adapting to climate change in pacific Island countries: The problem of uncertainty. *World Development, 29*(6), 977–993.

Batty, M. (2007). *Complexity in city systems: Understanding, evolution, and design.* Cambridge, MA: MIT Press.

Bettencourt, L., & West, G. (2010). A unified theory of urban living. *Nature, 467*(7318), 912–913.

Bonta, M., & Protevi, J. (2004). *Deleuze and geophilosophy: A guide and glossary.* Edinburgh: Edinburgh University Press.

Boyle, M., & Pogerson, R. J. (2001). Power, discourse and city trajectories. In R. Paddison (Ed.), *Handbook of urban studies* (pp. 402–425). London, UK: Sage Publications Ltd.

Bulkeley, H., Schroeder, H., Janda, K., Zhao, J., Armstrong, A., Chu, S. Y., & Ghosh, S. (2009). *Cities and climate change: The role of institutions, governance and urban planning.* Paper presented at the World Bank 5th Urban Symposium on Climate Change, June, Marseille.

Burby, R. J., Deyle, R. E., Godschalk, D. R., & Olshansky, R. B. (2000). Creating hazard resilient community through land-use planning. *Natural Hazards Review, 1*(2), 99–106.

Carpenter, S. R., Walker, B., Anderies, J. M., & Abel, N. (2001). From metaphor to measurement: Resilience of what to what? *Ecosystems, 4,* 765–781.

Castello, M. G. (2011). Brazilian policies on climate change: The missing link to cities. *Cities, 28* (6), 498–504.

CCC—Committee on Climate Change. (2010). *Building a low-carbon economy—the UK's innovation challenge.* www.theccc.org.uk.

CCC—Committee on Climate Change Adaptation. (2010). *How well prepared is the UK for climate change?* www.theccc.org.uk.

Chapin, III, F. S., Kofinas, G. P., & Folke, C. (Eds.). (2009). *Principles of ecosystem stewardship: Resilience-based natural resource management in a changing world.* New York: Springer Verlag.

Chapin, III, F. S., Power, M. E., Pickett, S. T. A., Freitag, A., Reynolds, J. A., Jackson, R. B., et al. (2011) Earth Stewardship: Science for action to sustain the human-earth system. *Ecosphere, 2* (8), 89.

Coaffee, J. (2006). From counter-terrorism to resilience. *European Legacy Journal of the International Society for the study of European Ideas, 11*(4), 389–403.

Coaffee, J. (2009). *Terrorism, risk and the global city: Towards urban resilience.* Famham: Ashgate Publishing.

Coaffee, J., & Rogers, P. (2008). Reputational risk and resiliency: The branding of security in place-making. *Place Branding and Public Diplomacy, 4,* 205–217.

Colten, C., Kates, R., & Laska, S. (2008). Three years after Katrina: Lessons for community resilience. *Environment: Science and Policy for Sustainable Development, 50,* 36–47.

Cooke, P., & Piccaluga, A. (Eds.). (2006). *Regional development in the knowledge economy.* NY: Routledge.

Corfee-Morlot, J., Kamal-Chaoui, L., Donovan, M. G., Cochran, I., Robert, A., & Teasdale, P. J. (2009). *Cities, climate change and multilevel governance.* OECD Environmental Working Papers 14: 2009, OECD publishing.

Cutter, S. L., Boruff, B. J., & Shirley, W. L. (2003). Social vulnerability to environmental hazards. *Social Science Quarterly, 84*(2), 242–261.

Cutter, S., Barnes, L., Berry, M., Burton, C., Evans, E., Tate, E., & Webb, J. (2008). A place-based model for understanding community resilience to natural disasters. *Global Environmental Change, 18,* 598–606.

Davic, R. D., & Welsh, H. H. (2004). On the ecological roles of salamanders. *Annual Review of Ecology Evolution and Systematics, 35*(1), 405–434.

De Roo, G., & Juotsiniemi, A. (2010). Planning and complexity. In *Book of Abstracts: 24th AESOP Annual conference.* Finland, p. 90.

Deleuze, G., & Guattari, F. (1991). *What Is philosophy?.* New York: Columbia University Press.

Dessai, S., & van der Sluijs, J. P. (2007). *Uncertainty and climate change adaptation—a scoping study.* Utrecht: Copernicus Institute for Sustainable Development and Innovation, Utrecht University.

Dodman, D., Hardoy, J., & Satterthwaite, D. (2009). *Urban development and intensive and extensive risk, background paper for the ISDR global assessment report on disaster risk reduction 2009.* London: International Institute for Environment and Development (IIED).

Eakin, H. C., & Wehbe, M. B. (2009). Linking local vulnerability to system sustainability in a resilience framework: Two cases from Latin America. *Climatic Change, 93,* 355–377.

Enfors, E. I., & Gordon, L. J. (2008). Dealing with drought: The challenge of using water system technologies to break dryland poverty traps. *Global Environmental Change, 18,* 607–616.

Ernstson, H., Barthel, S., Andersson, E., & Borgström, S. T. (2010). Scale-crossing brokers and network governance of urban ecosystem services: The case of Stockholm. *Ecology and Society, 15*(4), 28.

Folke, C. (2006). Resilience: The emergence of a perspective for social-ecological systems analyses. *Global Environmental Change, 16,* 253–267.

Folke, C., Carpenter, S. R., Walker, B. H., Scheffer, M., Chapin III, F. S., & Rockstro, J. (2010). Resilience thinking: Integrating resilience, adaptability and transformability. *Ecology and Society,* 15, 20. http://www.ecologyandsociety.org/vol15/iss4/art20/.

Folke, C., Jansson, Å., Rockström, J., Olsson, P., Carpenter, S. R., Chapin, F. S., et al. (2011). Reconnecting to the biosphere. *AMBIO: A Journal of the Human Environment, 40*(7), 719–738.

Fothergill, A., & Peek, L. (2004). Poverty and disasters in the United States: A review of recent sociological findings. *Natural Hazards, 32*(1), 89–110.

Godschalk, D. R. (2003). Urban hazards mitigation: Creating resilient cities. *Natural Hazards Review, 4*(3), 136–143.

Godschalk, D. R., Beatly, T., Berke, P., Brower, D. J., & Kaiser, E. J. (1999). *Natural hazard mitigation: Recasting disaster policy and planning.* Washington, DC: Island Press.

Gunder, M., & Hillier, J. (2009). *Planning in ten words or less: A Lacanian entanglement with spatial planning.* Farnham: Ashgate.

Gunderson, L., & Holling, C. S. (Eds.). (2001). *Panarchy: Understanding transformations in human and natural systems.* Washington, DC: Island Press.

Harriet, B. (2010). Cities and the governing of climate change. *Annual Review of Environment and Resources, 35,* 2.1–2.25.

Healey, P. (2007). *Urban complexity and spatial strategies: Towards a relational planning for our times.* New York: Routledge.

Healey, P., & Upton, R. (Eds.). (2010). *Crossing borders international exchange and planning practices.* Oxon: Routledge.

Hillier, J. (2010). Strategic navigation in an ocean of theoretical and practice complexity. In Hillier, J., & Healey, P. (Eds.), *The Ashgate research companion to planning theory: Conceptual challenges for spatial planning* (pp. 447–480). Farnham: Ashgate.

Holgate, C. (2007). Factors and actors in climate change mitigation: A tale of two South African cities. *Local Environment, 12,* 471–484.

Holling, C. (1973). Resilience and stability of ecological systems. *Annual Review of Ecology and Systematics, 4,* 1–23.

Jabareen, Y. (2006). Sustainable urban forms: Their typologies, models, and concepts. *Journal of Planning Education and Research, 26*(1), 38–52.

Jabareen, Y. (2009). Building conceptual framework: Philosophy, definitions and procedure. *International Journal of Qualitative Methods, 8*(4), 49–62.

Jameson, F. (1991). *Postmodernism or, the cultural logic of late capitalism.* Durham: Duke University Press.

Kaufman, E. (1998). Introduction. In E. Kaufman & K. J. Heller (Eds.), *Deleuze and Guattari: New mapping in politics, philosophy and culture* (pp. 3–19). Minneapolis, MN: University of Minnesota Press.

Kern, K., & Alber, G. (2008). Governing climate change in cities: Modes of urban climate governance in multi-level systems. In *Competitive Cities and Climate Change, OECD Conference Proceedings, Milan, Italy, 9–10 October 2008* (Chap. 8, pp. 171–196). Paris: OECD. http://www.oecd.org/dataoecd/54/63/42545036.pdf.

Leichenko, R. (2011). Climate change and urban resilience. *Current Opinion in Environmental Sustainability, 3*(3), 164–168.

Little, R. (2004). Holistic strategy for urban security. *Journal of Infrastructure Systems, 10*(2), 52–59.

Martin, R., & Sunley, P. (2007). Complexity thinking and evolutionary economic geography. *Journal of Economic Geography, 7*(4), 16–45.

Maru, Y. (2010). Resilient regions: Clarity of concepts and challenges to systemic measurement systemic measurement. In *Socio-economics and the environment discussion. CSIRO Working Paper Series.* http://www.csiro.au/files/files/pw5h.pdf.

Massey, D. (2003). Some times of space. In S. May (Ed.), *Olafur Eliasson: The weather project* (pp. 107–118). London: Tate Publishing. Exhibition catalogue.

McSweeney, K., & Coomes, O. (2011). Climate-related disaster opens 'window of opportunity' for rural poor in northeastern Honduras. *Proceedings of the National Academy of Sciences, USA, 108,* 5203–5208.

Morrow, B. H. (1999). Identifying and mapping community vulnerability. *Disasters, 23*(1), 1–18.

Nelson, A. C., & French, S. P. (2002). Plan quality and mitigating damage from natural disasters: A case study of the Northridge earthquake with planning policy considerations. *Journal of the American Planning Association, 68*(2), 194–207.

Newman, P., Beatley, T., & Boyer, H. (2009). *Resilient cities: Responding to peak oil and climate change.* Washington, DC: Island Press.

NYS (2013). *NYS 2100 commission: Recommendations to improve the strength and resilience of the empire state's infrastructure.*

Ojerio, R., Moseley, C., Lynn, K., & Bania, N. (2010). Limited involvement of socially vulnerable populations in federal programs to mitigate wildfire risk in Arizona. *Natural Hazards Review, 12*(1), 28–36.

Pais, J., & Elliot, J. (2008). Places as recovery machines: Vulnerability and neighborhood change after major hurricanes. *Social Forces, 86,* 1415–1453.

Pelling, M. (2003). *The vulnerability of cities: Natural disasters and social resilience.* London: Earthscan.

Pendall, R., Foster, K., & Cowel, M. (2010). Resilience and regions: Building understanding of the metaphor. *Cambridge Journal of Economic and Society, 3*(1), 71–84.

Pike, A., Dawley, S., & Tomaney, J. (2010). Resilience adaptation and adaptability. *Cambridge Journal of Regions, Economy and Society, 3,* 59–70.

Priemus, H., & Rietveld, P. (2009). Climate change, flood risk and spatial planning. *Built Environment, 35*(4), 425–431.

Protevi, J. (2006). Deleuze, guattari, and emergence. *Paragraph: A Journal of Modern Critical Theory, 29*(2), 19–39. http://www.protevi.com/john/Emergence.pdf.

Richardson, K. (Ed.) (2005). *Managing the complex vol. 1: Philosophy, theory and application.* Greenwich: Information Age Publishing. Bottom of Form.

Richardson, K. A., & Cilliers, P. (2001). What is complexity science? A view from different directions, *Emergence, 3*(1), 5–22.

Rodin, J., & Rohaytn, F. G. (2013). *NYS 2100 commission: Recommendations to improve the strength and resilience of the empire state's infrastructure.* http://www.governor.ny.gov/sites/governor.ny.gov/files/archive/assets/documents/NYS2100.pdf.

Romero Lankao, P. (2007). How do local governments in Mexico City manage global warming? *Local Environment, 12,* 519–535.

Rose, A. (2004). Defining and measuring economic resilience to disaster. *Disaster Prevention and Management, 13*(4), 307–314.

Roy, A. (2010). Informality and the politics of planning. In Hillier, J., & Healey, P. (Eds.), *Planning theory: Conceptual challenges for spatial planning* (pp. 87–107). Farnham: Ashgate Publishing.

Satterthwaite, D. (2008). *Climate change and urbanization: Effects and implications for urban governance.* Presented at UN Expert Group Meeting on Population Distribution, Urbanisation, Internal Migration and Development. UN/POP/EGMURB/2008/16/.

Schneider, S. H., Semenov, S., Patwardhan, A., Burton, I., Magadza, C. H. D., Oppenheimer, M., et al. (2007). Assessing key vulnerabilities and the risk from climate change. Climate change 2007: Impacts, adaptation and vulnerability. In M. L. Parry, O. F. Canziani, J. P. Palutikof, P. J. van der Linden, & C. E. Hanson (Eds.), *Contribution of working group II to the fourth assessment report of the intergovernmental panel on climate change* (pp. 779–810). Cambridge, UK: Cambridge University Press.

Simmie, J., & Martin, R. (2010). The economic resilience of regions: Towards an evolutionary approach. *Cambridge Journal of Regions, Economy and Society, 2010*(3), 27–43.

Turner, S. R. S., & Murray, M. S. (2001). Managing growth in a climate of urban diversity: South Florida's eastward ho! Initiative. *Journal of Planning Education and Research, 20,* 308–328.

Turok, I., & Mykhnenko, V. (2007). The trajectories of European cities, 1960–2005. *Cities, 24*(3), 165–182.

UNISDR-Inter-Agency secretariat of the International Strategy for Disaster Reduction (UN/ISDR) (2005). Building the resilience of nations and communities to disasters. In *Proceedings of the Conference: World Conference on Disaster Reduction (WCDR)*, United Nations, Geneva.

UNISDR-International Strategy for Disaster Reduction (2010). *Making cities resilient: My city is getting ready.* In 2010–2011 World Disaster Reduction Campaign.

United Nations Division for the Advancement of Women. (2001). *Environmental management and the mitigation of natural disasters: A gender perspective.* http://www.un.org/womenwatch/daw/csw/env_manage/documents/EGM-Turkey-final-report.pdf. July 7, 2009.

Urry, J. (2005). The complexity turn. *Theory, Culture and Society, 22*(5), 567–582.

Vale, J. L., & Campanella, T. J. (2005). *The resilient city: How modern cities recover from disaster.* New York: Oxford University Press.

Vellinga, P., Marinova, N. A., & van Loon-Steensma, J. M. (2009). Adaptation to climate change: A framework for analysis with examples from the Netherlands. *Built Environment, 35*(4), 452–470.

Walisser, B., Mueller, B., & McLean, C. (2005). *The resilient city. Prepared for the world urban forum 2006.* Canada: Vancouver Working Group, Ministry of Community, Aboriginal and Women's Services, Government of British Columbia.

Walker, W. E., Harremoës, P., Rotmans, J., van der Sluijs, J. P., van Asselt, M. B. A., Janssen, P., & Krayer von Krauss, M. P. (2003). Defining uncertainty: A conceptual basis for uncertainty management in model-based decision support. *Integrated Assessment, 4*(1), 5–17.

Walker, B. H., Anderies, J. M., Kinzig, A. P., & Ryan, P. (2006). Exploring resilience in social-ecological systems through comparative studies and theory development. *Ecology and Society,* 11, 12. http://www.ecologyandsociety.org/vol11/iss1/art12/.

Wardekker, J. A., de Jong, A., Knoop, J. M., & van der Sluijs, J. P. (2010). Operationalising a resilience approach to adapting an urban delta to uncertain climate changes. *Technological Forecasting and Social Change, 7*(6), 987–998.

WRI/WBCSD GHG Protocol. (2014). The global protocol for community-scale greenhouse gas emission inventories. http://ghgprotocol.org/files/ghgp/GHGP_GPC.pdf.

Zhang, Y. (2010). Residential housing choice in a multihazard environment: Implications for natural hazards mitigation and community environmental justice. *Journal of Planning Education and Research, 30*(2), 117–131.

Chapter 8
The Deficient Resilient Cities: Hurricane Sandy in New York City

8.1 Introduction

This chapter evaluates whether these urban plans and their policies are adequate, appropriate responses to environmental hazards and future climate change impacts. Toward this end, this chapter uses Hurricane Sandy—which drove a 4.2-m-high wall of salt water into the heart of New York City on 29 October 2012—as an example of the conditions that will manifest in cities around the world if the worst climate change scenarios are realized (NYS2100 Commission 2013; Tollefson 2012; Peltz and Hays 2012).

The resilience of contemporary cities to the anticipated impacts of climate change has become increasingly crucial to the wellbeing of urban residents, as we have witnessed in the devastating results and casualties of the environmental hazards that cities around the globe have faced recently. In recent years, many cities around the world, primarily in developed countries, have proposed urban plans and policies that are designed to increase their resilience to the anticipated, albeit uncertain, threats of climate change and environmental hazards. Urban planning is supposed to play a significant role in efforts to cope with climate change impacts in cities. Fleischhauer (2008) suggests that spatial planning can play an important role in mitigating multihazards by influencing urban structures and thus strengthening urban resilience.

The National Hurricane Center ranked Hurricane Sandy the second costliest US hurricane since 1900. Interestingly, climate change researchers suggest that sea levels in New York and elsewhere may rise by that same amount (4.2 m) by 2200. Even if Sandy was not caused by climate change, it provides a concrete illustration of one consequence of climate change (Chertoff 2012). Moreover, researchers suggest that this type of storm may be even stronger in the future, with fiercer winds and heavier rains (Plumer 2012). Hurricane Sandy has also generated renewed attention to the potential effects of climate change and the issue of city resilience. In fact, it provides a significant opportunity to examine the resilience of the affected

© Springer Science+Business Media Dordrecht 2015
Y. Jabareen, *The Risk City*, Lecture Notes in Energy 29,
DOI 10.1007/978-94-017-9768-9_8

cities, mainly those cities that have invested tremendous planning efforts in an attempt to counter climate change in particular (Gibbs and Holloway 2013: 1).

Furthermore, New York City has set forth an ambitious effort and begun to implement a broad and landmark plan, *PlaNYC 2030*, drawn up in 2007, to make the city more resilient to the expected effects of climate change (Solecki 2012: 570; Tollefson 2012; Rosenzweig and Solecki 2010b: 19; Rosan 2012; Jabareen 2014). However, the question at the heart of this chapter is whether these planning efforts improved NYC's ability to face this storm. This chapter aims to assess the planning policies that were designed to counter climate change impacts and the risk of environmental disasters in NYC, considering how NYC has prepared in recent years through planning and urban public policies targeting these issues. More specifically, this study assesses why the NYC planning policies were unable to adequately confront Hurricane Sandy. This chapter assesses the NYC planning policies and analyzes why it was unable to adequately confront Hurricane Sandy.

8.2 Assessment Method: The Contribution of Planning to City Resilience

The previous chapter discusses the concept of resilience and suggests that city resilience is a complex and multidisciplinary phenomenon. Generally, the literature on measuring resilience overlooks the contribution of urban planning in general and of specific plans to the resilience of cities and communities. The literature has instead tended to focus on ecosystems and to suggest quantitative indicators for such assessments. The existing frameworks assisted researchers in assessing the resilience of cities but are less helpful in assessments of specifically urban plans because they overlook the spatial dimensions of planning.

The problem with resilience measurement is to identify what is an acceptable level of loss would be, and what an acceptable time to recovery would be, in order to qualify as 'resilient. What is the 'operational definition' or the operationalization process of defining the measurements? For example, what is the acceptable cost of an environmental extreme event? How much is unacceptable? What is the unaccepted time of recovery? These and other questions remain almost without adequate answers. This issue needs to be discussed jointly by the public, and the academic and institutional governmental levels.

In order to assess the contribution of planning efforts to the resilience of city, this chapter suggests a method that is based on two types of criteria, which will be examined in order to assess the resilience of the city aftermath of Hurricane Sandy:

1. The first is related to the recovery and costs of the storm and includes three criteria:

   (a) *Recovery Time*: What was the duration recovery time of the city? In order to operationalize and quantify this criterion, I suggest that a duration recovery time of the essential services and infrastructure, mainly main transportation modes, electricity and water, should be 48 h. This is in order not to risk lives of people.

(b) *Casualties*: What were the tolls of the event? The city should accept zero-casualty. This is the working objective of any plans for city resilience.

(c) *Costs of event*: What is the scope of damages to society, urban infrastructures, and economy? This type of cost is not easy to anticipate and of course it should be as minimum as we can plan and act.

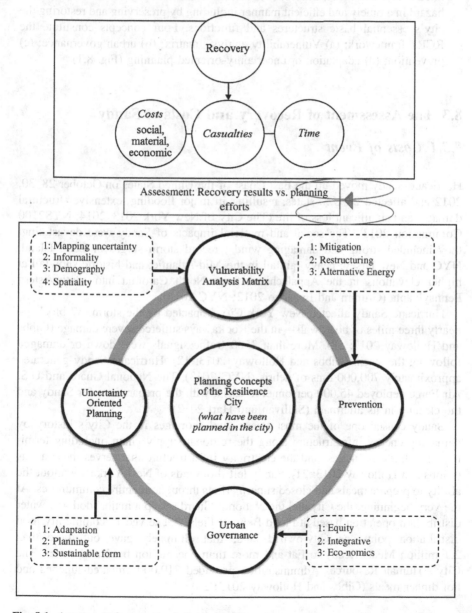

Fig. 8.1 Assessment framework of city resilience

2. The second is related to the framework of the resilience city and its planning concepts, which is presented in the previous chapter and composed. The previous chapter suggests the *Risk City Resilience Trajectories*,' or RCRT, which addresses the question of what cities and their urban communities should do in terms of planning to move towards a more resilient state in the future. According to the RCRT, a resilient city is defined by the overall ability of its governmental, physical, economic and social systems and entities to learn, be prepared, plan for uncertainty, and resist, absorb, accommodate and recover from the effects of a hazard in a timely and efficient manner including by preserving and restoring the city's essential basic structures and functions. Four concepts constitute the RCRT framework: (a) Vulnerability analysis matrix; (b) urban governance; (c) prevention; (d) adaptation or uncertainty-oriented planning (Fig. 8.1).

8.3 The Assessment of Recovery and Costs of Sandy

8.3.1 Costs of Event

Hurricane Sandy moved up the East Coast of the United States on October 28–30, 2012 and affected 24 U.S. states, resulting in major flooding, extensive structural damage, and significant loss of life (The City of New York 2013, 2014; NYS2100 Commission 2013). The social and material impacts of Sandy were devastating; they included strong and damaging winds; record storm surges in Long Island, NYC and New Jersey; heavy rainfall in the Mid-Atlantic; and historic snowfall at higher elevations in the Appalachians from North Carolina into southwestern Pennsylvania (Grumm and Evanego 2012; NYC 2013).

Hurricane Sandy affected New York City impacted by the storm. 37 blocks—nearly three miles of boardwalk—in the Rockaways suffered severe damage (Gibbs and Holloway 2013: 18). More than 3500 traffic signals were down or damaged following the storm (Gibbs and Holloway 2013: 12). Hurricane Sandy generated approximately 700,000 tons of debris (NYC 2013). The National Guard and U.S. Air Force deployed 45,000 personnel to assist with the preparations for Sandy and the cleanup in its aftermath (Sullivan and Hart 2012).

Sandy caused one of the most severe fuel shortages in the City's history by damaging energy infrastructure along the regional supply chain, including terminals, pipelines, refineries, and the electricity infrastructure that serves these assets (Gibbs and Holloway 2013: 21). Sandy left thousands of New Yorkers without the ability to prepare meals and closed supermarkets throughout entire communities. At the very beginning, the City and the National Guard set up a major food and water distribution operation based at Floyd Bennett Field that served 17 community food distribution points on City-owned land, that ultimately gave out more than 2.1 million Meals Ready to Eat and more than one million bottles of water. The City's Human Resources Administration provided 719,000 prepared lunches and hot dinner meals (Gibbs and Holloway 2013: 26).

About 23,400 citywide businesses with approximately 245,000 employees are located in flood-impacted areas, 95 % of which are small- and medium sized enterprises that employ 50 people or fewer. These businesses faced extensive damage from loss of inventory, ruined equipment, damage to the interior of their space, and structural damage to building systems (Gibbs and Holloway 2013: 30).

In early December 2012, President Obama signed the executive order that established the Hurricane Sandy Rebuilding Task Force. President Obama also asked Congress to immediately approve $60 billion in supplemental assistance to aid in the storm recovery efforts. The main aim of the Task Force, led by the Federal Emergency Management Agency (FEMA) within the Department of Homeland Security, is leading the recovery efforts to assist the affected region. The Executive Order, entitled Establishing the Hurricane Sandy Rebuilding Task Force, states that "Rebuilding efforts must address economic conditions and the region's aged infrastructure—including its public housing, transportation systems, and utilities—and identify the requirements and resources necessary to bring these systems to a more resilient condition given both current and future risks." On January 29, 2013, President Obama signed into law the "Disaster Relief Appropriations Act, 2013" (Public Law 113-2), which provides $16 billion in Community Development Block Grant Disaster Recovery (CDBG-DR) funds to repair and restore areas affected by Hurricane Sandy. NYC's first round of CDBG-DR funds is $1.77 billion. The City developed a partial Action Plan ("Action Plan A") that details how it will use this funding to help New Yorkers rebuild their homes, businesses and communities (Gibbs and Holloway 2013: 34).

Goldstein et al. (2014: 2), the Senior Advisor for Recovery, Resiliency and Infrastructure of NYC, contends that "Hurricane Sandy highlighted New York City's vulnerability to extreme weather events; a vulnerability that will only grow over time due to the impacts of a changing climate". Eventually, Sandy "caused more than $19 billion in damage and lost economic activity, thousands of homes and businesses were destroyed or seriously impacted, infrastructure systems and vital services that serve millions were disrupted, and 44 New Yorkers tragically lost their lives. Sandy also exposed other underlying challenges in many neighborhoods, where many of the city's most vulnerable populations live, and where individuals and families are at even higher risk of disruption, dislocation, and displacement" (Goldstein et al. 2014: 2).

8.3.2 Casualties

The hurricane killed more than one hundred individuals among them 34 New Yorkers (Gibbs and Holloway 2013: 1). This is an extremely high toll that cities should not be allowed.

8.3.3 Recovery Time

The recovery of New York City from Sandy is still ongoing and it "will continue as long as there are New Yorkers who are displaced from their homes and businesses, and until neighborhoods have fully recovered from the storm. At the time of this report the NYPD still maintains a contingent of more than 125 officers assigned to storm-affected areas 24 h a day" (Gibbs and Holloway 2013: 32).

Thousands of families in the hardest hit areas did not have power restored for months following the storm (NYC 2013). Health, food and water distribution, and distribution sites and Restoration Centers helped in addressing the needs of many New Yorkers as the NYC suggested (Gibbs and Holloway 2013: 27). In addition to providing a safe home for New Yorkers through Rapid Repairs, the City launched a set of programs to help businesses recover from both physical damage and losses from extended closures.

Hurricane Sandy resulted in a need for 150,000 New Yorkers for temporary housing or immediate home repairs (Gibbs and Holloway 2013: 39). Yet, "for those evacuees who remained in emergency shelters but were unable to return to their homes, the City entered into agreements with hotels to provide alternative stable, short-term evacuation sheltering" (Gibbs and Holloway 2013: 39).

A Post Sandy Survey for Zone A, which was conducted for NYC, which randomly sampled 509 interviews (margin of error ±4.3 %), found that 56 % of the interviewees "lost power for more than one week", 55 % of them "lost other services", and 43 % mentioned that their "home was damaged" (Gibbs and Holloway 2013; Appendix B).

The City of New York report *One City, Rebuilding Together* (Goldstein et al. 2014: 6) suggests that within two weeks of the storm the city launched Rapid Repairs, "a first-of-its-kind emergency sheltering program to provide essential repairs to thousands of homeowners left without heat, power, or hot water following Hurricane Sandy" and that "in less than 100 days, Rapid Repairs restored heat, power and hot water service to over 11,700 buildings—which included over 20,000 units—and addressed the needs of approximately 54,000 New Yorkers. The total cost of the Rapid Repairs program is estimated at approximately $640 million, over $604 million of which has already been paid out for direct construction costs and indirect program costs." The help of NYC recovery, as shown in Table 8.1, has come primarily through the federal government's CDBG-DR grant, which is administered by the U.S. Department of Housing and Urban Development. CDBG-DR grants are resources allocated to help areas recover from presidentially declared disasters. In mid-January 2013, Congress passed the Disaster Relief Appropriations Act, which was the legislative vehicle for distributing CDBG-DR grants to areas impacted by Hurricane Sandy (Goldstein et al. 2014: 6).

In sum, the human, social, and material costs of Sandy were huge. The recovery, which is still ongoing, took a week in order to restore power to most part of the city. However, there still lies a large challenge ahead for full recovery (Table 8.2).

Table 8.1 The assessment framework for planning the resilient city

Concepts	Components	Key questions (measurements)
Criteria related to the costs of the storm		
Recovery	C1: Time	*Time*: what was the duration recovery time of the city?
	C2: Casualties	*Casualties*: what were the tolls of the event?
	C3: Cost	*Costs of event*: what is the scope of damages to the urban social, material and economy infrastructures?
Criteria related to the planning and plans		
Concept 1: urban vulnerability matrix analysis	C1: Uncertainties	C1. What are the hazard and environmental uncertainties?
	C2: Informality	C2. What is the scope, geography, socio-economic, demographic, and physical characters of existing informal settlements in or closed to the city?
	C3: Demography	C3. What is the nature of vulnerable demography in the city by age, gender, health, and other social group?
	C4: Spatiality	C4. What is the spatial distribution of environmental hazards and risks?
Concept 2: uncertainty oriented planning	C1: Adaptation	C1: What adaptation measures are taken to reduce risks and cope with future uncertainties?
	C2: Planning	C2: How do planning methods cope with uncertainties?
	C3: Sustainable form	C3: What are characteristics of the existing and planned urban form typologies?
Concept 3: urban governance	C1: Equity	C1: Who participates in decision-making and planning regarding environmental and uncertainty issues?
	C2: Integrative	C2: Is the urban governance approach integrating institutional, legal, social, economic, and environmental aspects?
	C3: Eco-economics	C3: What is the nature of the existing and planned ecological economy?
Concept 4: prevention	C1: Mitigation	C1: What mitigation measures are taken to reduce risks and to prepare the city for future environmental hazards?
	C2: Restructuring	C2: What are the proposed or planned spatial, physical, and economic restructuring policies that aim to face the environmental hazards and uncertainties?
	C3: Alternative energy	C3: How does the city address the energy sector and whether it proposes strategies to reduce energy consumption and to use new alternative and cleaner energy sources?

Table 8.2 CDBG-DR allocations for NYC recovery

CDBG-DR funds allocated to NYC and the current allocation program name	Total allocations
Housing programs	$1,695,000,000 (52.6 %)
Build it back rehabilitation and reconstruction	$1,022,000,000
Build it back multifamily building	$346,000,000
Rental assistance	$19,000,000
Public housing rehabilitation and resilience	$308,000,000
Business programs	$266,000,000 (8.3 %)
Business loan and grant program	$42,000,000
Business resiliency investment program	$110,000,000
Neighborhood game changer investment	$84,000,000
Resiliency innovations for a stronger economy	$30,000,000
Infrastructure and other city services	$805,000,000 (25.0 %)
Public services	$367,000,000
Emergency demolition	$2,000,000
Debris removal/clearance	$12,500,000
Code enforcement	$1,000,000
Rehabilitation/reconstruction of public facilities	$324,500,000
Interim assistance	$98,000,000
Resiliency	$284,000,000 (8.8 %)
Coastal protection	$224,000,000
Residential building mitigation program	$60,000,000
Citywide administration and planning	$169,820,000 (5.3 %)
Planning	$72,820,000
Administration	$97,000,000
Total	*$3,219,820,000 (100 %)*

Source Goldstein et al. (2014: 8): *one city, rebuilding together*

8.4 Climate Change-Oriented Planning in New York City

NYC initiated a new plan for the city, *PlaNYC 2030: A Greener, Greater New York*, on Earth Day 2007. This plan was presented in a previous chapter; therefore, here I will focus on the themes that are related to planning the resilience city. *PlaNYC* identifies three primary challenges for the plan to address: growth, an aging infrastructure, and an increasingly precarious environment (PlaNYC: 4). Climate change is a primary factor in the problems facing NYC and helped to explain the urgency of the new plan. *PlaNYC 2030* is composed of 127 new initiatives that aim to strengthen the economy, public health, and quality of life of the city, and "they will also form a frontal assault on the biggest challenge of all: global climate change" (The City of New York 2009: 2see also The City of New York, 2014). Collectively, these initiatives are intended to achieve a 30 % reduction in GHG emissions by 2017 (Inventory of NYC Greenhouse Gas Emission 2009).

As mentioned, four concepts constitute the RCRT framework which will be assessed in the following section: (a) Vulnerability analysis matrix; (b) urban governance; (c) prevention; (d) adaptation or uncertainty-oriented planning.

8.4.1 Vulnerability Analysis Matrix

Unlike conventional plans, *PlaNYC* makes the climate change issue the heart of its mission and its point of departure. In its introduction, *PlaNYC* identifies the expected future climate change impacts and the relevant scenario uncertainty. A fundamental assumption of *PlaNYC* is that "climate change poses real and significant risks to New York City" (PlaNYC: Progress Report, 2009: 39). *PlaNYC* portrays New York as a city at risk. Therefore, *PlaNYC's* vision generates a sense of local and global *urgency*: "unless the public...appreciate[s] the urgency," the plan warns, "we will not meet our goal" (PlaNYC: 110). Ironically, PlaNYC acknowledges: "Meanwhile, we will face an increasingly precarious environment and the growing danger of climate change that imperils not just our city, but the planet. We have offered a different vision" (PlaNYC: 141).

PlaNYC acknowledges that shifting climate patterns will have a wide range of effects on these communities, taking lives, posing "major public health dangers," and affecting the property and livelihoods of many (PlaNYC: 138). Yet, *PlaNYC* fails to address how climate change could affect each neighborhood and to emphasize the specific environmental risks that exist in each neighborhood and that each neighborhood is likely to face in the future. NYC is a diverse city with 5 boroughs, 59 community districts and hundreds of neighborhoods. Moreover, the estimated population of NYC at the end of 2012 was 8,244,910 people speaking 174 different languages (US Census Bureau 2013). All five NYC boroughs "have vulnerable coastline." Moreover, the massive growth envisioned by *PlaNYC* will certainly affect these communities and may even "erase the character of communities across the city" (PlaNYC: 18). In considering the spatial impact of implementing the plan, the planners raise a crucial dilemma for the future of NYC and its communities: "We cannot simply create as much capacity as possible; we must carefully consider the kind of city we want to become. We must ask which neighborhoods would suffer from the additional density and which ones would mature with an infusion of people, jobs, stores and transit. We must weigh the consequences of carbon emissions, air quality, and energy efficiency when we decide the patterns that will shape our city over the coming decades" (PlaNYC: 18).

8.4.2 Urban Governance

In the context of urban governance, *PlaNYC* failed to approach various communities and let the people and neighborhoods participate in its designing and forging critical strategies for their spaces.

NYC is portrayed as a city at risk, and therefore, it is suggested that the City needs "to rethink the way it operates and adapts to its evolving environment" (NPCC 2009: 3). The plan's primary strategy for climate change adaptation appears to lie in the creation of "an intergovernmental Task Force to protect our city's vital infrastructure" and "to work with vulnerable neighborhoods to develop site-specific strategies" (PlaNYC: 136). *PlaNYC* proposes the establishment of a NYC Climate Change Advisory Board, a citywide strategic planning process, "to determine the impacts of climate change to public health and other elements of the City and begin identifying viable adaptation strategies" (PlaNYC: Progress Report 2009: 39). Without a doubt, these bodies, which are charged with monitoring climate change parameters vis-à-vis the city and proposing adjustment policies, enhance the city's urban adaptive planning capacity. However, the plan's adaptation strategy is also principally based on emissions reduction, an *ex-ante* strategy. In this way, *PlaNYC* fails to prepare the city and its infrastructure for the disasters that could stem from climate change. For example, the plan proposes no infrastructure design or development projects along the city's vulnerable 570 miles of coastal zones. In contrast, *PlaNYC* proposes to intensify development wherever possible in waterfront and other areas without considering the risks posed by climate change. Lastly, *PlaNYC* proposes no *ex-post* strategy or emergency response to such disasters.

8.4.3 Prevention: Mitigation

PlaNYC promotes mitigation initiatives in order to contribute its part in preventing or easing climate change impact. These initiatives are intended to improve air quality and reduce emissions by 30 % by 2030 as mentioned in a previous chapter. By 2030, at least 85 % of the city's energy will be used by currently existing buildings. In this way, energy saving measures in existing buildings will result in a seven million ton reduction in global warming emissions. This reduction is significant because without the measures outlined in the plan, emissions would rise to nearly 80 million metric tons by 2030 (PlaNYC: Progress Report 2009: 39).

8.4.4 Adaptation: Uncertainty-Oriented Planning

There are various challenges related to climate change threats. In addition, the climate change threats to NY are exacerbated by the already deteriorating physical condition of the city's infrastructure, which dramatically contributes to the uncertainties surrounding climate change.

Rosenzweig and Solecki (2010a, b) suggest that current climate change adaptation efforts in New York City have been conducted in the last decade by various leading agencies, including the New York City Department of Environmental Protection (NYCDEP), the Port Authority of New York and New Jersey, and nongovernmental organizations such as the Environmental Defense Fund. In 2004, the NYCDEP, which is responsible for the New York City water and waste water systems and its drain water, launched the Climate Change Task Force initiative with the mission of examining the potential risks of climate change to the city's water supply, drainage, and wastewater management systems and of integrating GHG emissions management to the greatest extent possible (Rosenzweig and Solecki 2010a, b). The major product of the NYC DEP Task Force was the Climate Change Assessment and Action Plan for the agency (NYC DEP: 2008).

Regarding adaptation policies, *PlaNYC* asserts that "there is no silver bullet to deal with climate change," and "as a result, our strategy to help stem climate change is the sum of all the initiatives in this plan" (PlaNYC: 135). The plan's primary climate change adaptation strategy appears to be the creation of "an intergovernmental Task Force to protect our city's vital infrastructure" and "to work with vulnerable neighborhoods to develop site-specific strategies" (PlaNYC: 136). In addition, *PlaNYC* proposes the establishment of a NYC Climate Change Advisory Board and a citywide strategic planning process "to determine the impacts of climate change to public health and other elements of the City and begin identifying viable adaptation strategies" (PlaNYC: Progress Report 2009: 39). The proposed adaptation policies also include measures intended to fortify the city's critical infrastructure, which are to be implemented through close cooperation among city, state, and federal agencies and authorities; updates to the flood plain maps to better protect the areas that are most vulnerable to flooding; and work with at-risk neighborhoods across the city to develop site-specific plans. "In addition to these targeted initiatives," the plan reads, "we must also embrace a broader perspective, tracking the emerging data on climate change and its potential impacts on our city" (PlaNYC: 136).

Moreover, the NPCC proposes a multistep adaptation planning process that includes identifying climate hazards and impacts, developing and evaluating adaptation strategies, implementing actions, and monitoring results to incorporate climate change adaptation planning into its existing planning and operational processes (Rosenzweig and Solecki 2010a, b: 14).

Without a doubt, the devastating effects of Sandy on NYC suggest that the city's planning efforts, mainly the recent *PlaNYC*, failed to protect it. It appears that NYC lacks the resiliency and adaptation that are necessary to survive such environmental hazards without significant damage.

A Costal Storm Plan, as collection of programs to prepare and respond to a storm, including evacuation, sheltering and logistic planning was prepared in 2000. In 2000, New York City first released its citywide plan for hurricanes, and in 2006, the Coastal Storm Plan (CSP) was updated to account for New York City's

Table 8.3 Resilience assessment of *PlaNYC*

Concepts/criteria	Achievements	Shortcomings	Achievement level
Recovery			
Time	A bout one week for power restoration	Still ongoing recovery	Moderate
Casualties		34 New Yorkers were killed	Extremely weak
Costs		Huge social, economic, and physical costs	Extremely weak
Planning and resilience			
Urban vulnerability matrix analysis	1. Analyzing the impacts of climate change, related uncertainty and environmental hazards on the city	1. *PlaNYC* does not include an extensive analysis of the spatial and demographic distribution of vulnerability and risk in the city	Moderate
Prevention	1. Proposing various and multifaceted *mitigation* measures to reduce greenhouse gas emissions		Good
	2. Addressing the energy sector and proposing strategies for reducing energy consumption and for the use of new, alternative and cleaner energy sources		
Adaptation: uncertainty oriented planning		1. Very few adaptation measures are proposed as a means to reduce risks and cope with future uncertainty, and few policies are proposed as a means to reduce risk and prepare the city for future environmental hazards	Extremely weak
		2. There is a lack of spatial and physical restructuring policies and projects intended to help the city cope with environmental hazards	
Urban governance	1. Proposing a new institutional and formal framework for coping with climate change impacts and achieving sustainability	1. There was a lack of widespread public participation in preparing *PlaNYC*	Moderate

changing population and to take into account the lessons learned from Hurricane Katrina. The CSP delineates three evacuation zones, A, B, and C. Zone A includes the City's coastline and low-lying areas most vulnerable to a costal storm. Thousands of people did not leave the evacuation zone, and 43 New Yorkers lost their lives.

The *NYS2100 Commission,* which was convened by Governor Andrew Cuomo in response to the unprecedented, recent, and severe weather events experienced by New York State and the surrounding region—Superstorm Sandy, Hurricane Irene, and Tropical Storm Lee—acknowledges the fragile resilience level of NYC and NYS. The Commission reviewed the vulnerabilities faced by the State's infrastructure systems and developed specific recommendations that they suggested should be implemented to increase New York's resilience in five main areas: transportation, energy, land use, insurance, and infrastructure finance. The *NYS2100 Commission* (2013) suggests that: "We cannot prevent all future disasters from occurring, but we can prevent failing catastrophically by embracing, practicing, and improving a comprehensive resilience strategy. As New York and our neighboring states continue to recover from the devastating impacts of Superstorm Sandy, we have a narrow but distinct window of opportunity to leverage the groundswell of consciousness. The next century will be defined by the extent to which our communities are resilient to the direct and indirect impacts of a rapidly changing climate and other long-term accelerators of change" (2013: 7, 10).

The plan suggests problematic mega projects around the city's waterfront. NYC has 570 miles of waterfront, which the Plan regards as "one of the city's greatest opportunities for residential development" and an important site for other types of projects (PlaNYC: 22). *PlaNYC* also confronts the "legacy of the City's industrial past..." "...which treated New York's waterways as a delivery system" (PlaNYC: 51) and proposes the opening of 90 % of the city's waterways to recreation by preserving natural areas and reducing pollution (PlaNYC: 53) (Table 8.3).

8.5 Conclusions

The devastating effects of Hurricane Sandy, like many other environmental hazards that have confronted numerous cities around the world, provide a significant opportunity to examine the resilience level of the city and to determine how the city should act in the future to confront the impacts and hazards of climate change. This chapter concludes that, although the City of New York has a plan for countering climate change impacts and although it has begun implementing its plan's projects, the city appears to be unable to cope with future serious climate impacts. Table 8.1 summarizes the achievements of *PlaNYC* and the city's efforts to counter and cope with climate change impacts in general as well as the city's resilience level in facing Hurricane Sandy. Apparently, as the impacts of Hurricane Sandy revealed, the major, critical shortage of the plan is its *adaptation* measures for coping with environmental hazards.

As the *NYS2100 Commission* (2013) concludes: "Superstorm Sandy produced countless stories of heartbreak but also of hope and resilience" (2013: 7). Yet, NYC, like many cities around the world, including the most pioneering among them, still fails to use comprehensive and spatial planning in its fight against climate change (Kern and Alber 2008). Barnett (2001) correctly argues that adaptation is difficult to grasp because it demands system-wide analysis and intervention. Most cities appear to be using mitigation policies to address human sources of climate change by reducing greenhouse gas emissions but have failed to apply adaptation policies. Wheeler (2008) suggests that the "first generation" of local climate action plans dealt overwhelming with mitigation rather than adaptation policies. Baker et al. (2012) suggest that although cities were aware of expected climate change impacts, their capacity to use this information to develop geographically specific action plans was limited. In their empirical study, Baker et al. (2012) found that none of the local adaptation plans that they examine provided comprehensive coverage of all of the outcome criteria in relation to the plan components. Moreover, preliminary research in developed countries indicates that implementing effective local adaptation plans may be beyond the capacity of many local governments (Wilson 2006).

Unfortunately, Sandy reveals that the current institutional and spatial settings of our cities are not resilient and that our cities become risky places for their residents during hazardous events. Uken (2012) suggests that the storm highlights the fragility of the aging American infrastructure, with an electricity network that is ranked below those of considerably poorer nations. Doubtless, NYC, like the rest of the US cities affected by Sandy, was not resilient in confronting it. Thus, the question of whether such cities are capable of facing the future challenges of climate change without implementing dramatic adaptation measures and other related policies must be raised. At present, NYC and other US cities appear unable to cope with environmental hazards. Moreover, researchers suggest that due to global warming, the number of future hurricanes will "either decrease or remain essentially unchanged" overall, but those that do form, as in the area of NY, will likely be stronger, with fiercer winds and heavier rains (Plumer 2012). If NYC is unprepared to face these extreme hazards, its residents and the rest of the city's systems will be devastatingly harmed.

The Commission of the European Communities (CEC 2009: 3) suggests that "even if the world succeeds in limiting and then reducing GHG emissions, our planet will take time to recover from the greenhouse gases already in the atmosphere. Thus, we will be faced with the impact of climate change for at least the next 50 years. We need therefore to take measures to adapt." The CEC (2009) regrets the "piecemeal manner" in which adaptation policies have been implemented and concludes that "a more strategic approach is needed to ensure that timely and effective adaptation measures are taken, ensuring coherency across different sectors and levels of governance" (CEC 2009: 3).

NYC did not have adequate public participation processes (Jabareen 2013; Jabareen 2014), which contributed to the low level of resilience of the city and of its vulnerable neighborhoods and areas. Baker et al. (2012) suggest that the local governments had not effectively planned for climate impacts and that local governments need to genuinely engage in public participation programs when developing climate adaptation plans. Effective public participation is an essential element of the planning process (Preston et al. 2011; Wiseman et al. 2010) and will help to balance the requirement for more standardized plans by ensuring that local communities participate in the framing of the climate adaptation problem, which is inherently localized and contextual in nature (Baker et al. 2012). Moreover, plans developed in partnership with communities are also more likely to be implemented (Wiseman et al. 2010).

In the aftermath of Sandy, NYC and NYS were aware of the extent of the disaster and the need for adaptation policies and strategies (NYC 2013; NYS2100 Commission 2013). At the city level, NYC (NYC 2013) suggests a roadmap of strategic steps that the city will take to improve its ability to protect life and property in the face of the increasing risk of severe weather and to strengthen the city's overall preparedness and generate the building blocks for a thorough and organized response to extended emergency events that may impact thousands of New Yorkers (2013: 5). In addition, the *NYS2100 Commission* (2013), which has also responded to Sandy, has suggested certain adaptation strategies.

The critical task of the city at the present is to prepare it for the uncertain futures. It is important to learn and use the crisis after Sandy in order to build the futures of the city and promote its resilience for all. Convincingly, Judith Robin, the President of Rockefeller Foundation and Felix Rohatyn state (NYS2100 Commission 2013: 7):

As New York continues to recover, we must also turn our attention to the future. We live in a world of increasing volatility, where natural disasters that were once anticipated to occur every century now strike with alarming regularity. Our response capabilities to this new level of instability and the ability to bounce back stronger must be developed and strengthened. Our efforts must be rooted in robust structural underpinnings as well as expanded operational capacities. Superstorm Sandy made the urgency of this undertaking painfully clear. We also now possess a vastly deeper understanding of our current vulnerabilities. We cannot just restore what was there before—we have to build back better and smarter. As Governor Cuomo said, 'It's not going to be about tinkering on the edges. Many of these systems we know have not worked for many, many years.'

References

Baker, I., Peterson, A., Brown, G., & McAlpinea, C. (2012). Local government response to the impacts of climate change: An evaluation of local climate adaptation plans. *Landscape and Urban Planning, 107*, 127–136.

Barnett, J. (2001). Adapting to climate change in pacific island countries: The problem of uncertainty. *World Development, 29*(6), 977–993.

CEC—The Commission of the European Communities. (2009). *White paper: Adapting to climate change: Towards a European framework for action.* Brussels.

Chertoff, E. (2012). The sandy storm surge: Is this what climate change will look like? *The Atlantic.* October 30, 2012.

Fleischhauer, M. (2008). The role of spatial planning in strengthening urban resilience. In H. J. Pasman (Ed.), *Resilience of cities to terrorist and other threats. NATO Science for Peace and Security Series Series C: Environmental Security* 2008 (pp. 273–298).

Gibbs, L., & Holloway, C. (2013). *Hurricane sandy after action: Report and recommendations to Mayer Michael R.* New York City: Bloomberg.

Goldstein, W., Peterson, A., & Zarrilli, D. A. (2014). *One city, rebuilding together: A report on the city of New York's response to hurricane sandy and the path forward.* New York City. http://www1.nyc.gov/assets/home/downloads/pdf/reports/2014/sandy_041714.pdf.

Grumm, R. H., & Evanego, C. (2012). *Hurricane sandy: An eastern United States superstorm-draft.* PA: National Weather Service State College.

Jabareen, Y. (2013). Planning for countering climate change: Lessons from the recent plan of New York city—PlaNYC 2030. *International Planning Studies, 18*(2), 221–242.

Jabareen, Y. (2014). An assessment framework for cities coping with climate change: The case of New York city and its PlaNYC 2030. *Sustainability, 6*(9), 5898–5919.

Kern, K., & Alber, G. (2008). Governing climate change in cities: modes of urban climate governance in multi-level systems. In *OECD Conference Proceedings Competitive Cities and Climate Change* (Chapter 8) (pp. 171–196). Milan, Paris, Italy: OECD. October 9–10, 2008. http://www.oecd.org/dataoecd/54/63/42545036.pdf.

NYS. (2013). NYS2100 commission: Recommendations to improve the strength and resilience of the empire state's infrastructure.

Pais, J., & Elliot, J. (2008). Places as recovery machines: Vulnerability and neighborhood change after major hurricanes. *Social Forces, 86*, 1415–1453.

Peltz, J., & Hays, T. (2012). Hurricane sandy: Storm surge floods NYC tunnels, cuts power to city. *The Christian Science Monitor.* October 29, 2012.

Plumer, B. (2012). Is Sandy the second-most destructive U.S. hurricane ever? Or not even top 10? *The Washington Post.* November 5, 2012.

Preston, B., Westaway, R., & Yuen, E. (2011). Climate adaptation planning in practice: An evaluation of adaptation plans from three developed nations. *Mitigation and Adaptation Strategies for Global Change, 16*(4), 407–438.

Rosan, C. D. (2012). Can PlaNYC make New York City "greener and greater" for everyone?: Sustainability planning and the promise of environmental justice. *Local Environment, 17*(9), 959–976.

Rosenzweig, C., & Solecki, W. (2010a). Introduction to climate change adaptation in New York city: Building a risk management response. *Annals of the New York Academy of Sciences, 1196*, 13–17 (Issue: New York City Panel on Climate Change 2010 Report).

Rosenzweig, C., & Solecki, W. (2010b). Chapter 1: New York city adaptation in context. *Annals of the New York Academy of Sciences* (Issue: New York City Panel on Climate Change 2010 Report).

Solecki, W. (2012). Urban environmental challenges and climate change action in New York City. *Environment and Urbanization, 24*, 557–573.

Sullivan, B. K., & Hart, D. (2012). Hurricane sandy barrels northward, may hit New Jersey (pp. 10–28). http://www.bloomberg.com/news/.

The City of New York. (2009). PlaNYC Progress Report 2009. The City of New York. Available from http://www.nyc.gov.

The City of New York. (2013). PlaNYC Progress Report 2013. The City of New York. Available from http://www.nyc.gov.

The City of New York. (2014). PlaNYC Progress Report 2014. The City of New York. Available from http://www.nyc.gov/html.

Tollefson, J. (2012). Hurricane sweeps US into climate-adaptation debate. *Nature, 491*, 167–168.

Uken, M. (2012). Sandy zeigt, wie marode Amerikas Infrastruktur ist [Sandy shows how ailing America's infrastructure is] (in German). *Zeit Online* (Hamburg, Germany), pp. 10–30. Retrieved November 02, 2012.

US Census Bureau. (2013). US Census 2013. Available from www.census.gov.

Wheeler, S. M. (2008). State and municipal climate change plans—the first generation. *Journal of the American Planning Association, 74*(4), 481–496.

Wilson, E. (2006). Adapting to climate change at the local level: The spatial planning response. *Local Environment, 11*, 609–625.

Wiseman, J., Williamson, L., & Fritze, J. (2010). Community engagement and climate change: Learning from recent Australian experience. *International Journal of Climate Change Strategies and Management, 2*(2), 134–147.

Chapter 9
The Inequality of the Risk City: Socio-Spatial Mapping the Risk City

9.1 Introduction

Analyzing and understanding future vulnerabilities are significant for the *risk city* and its planning practices that aim to cope with them. Vulnerabilities may be related to social, economic, environmental and climate change risk. In this chapter I will deal only with climate change and environmental vulnerabilities. Accordingly, vulnerability is the "degree to which a system is susceptible to, and unable to cope with, adverse effects of climate change, including climate variability and extremes. Vulnerability is a function of a system's exposure, its sensitivity, and its adaptive capacity" (CCC 2010: 61). Vulnerability assesses risk in relation to multiple and interacting stresses (McLaughlin and Dietz 2008; Nelson 2012). It assesses the attributes of sensitivity, exposure and the adaptation capacity of a group (Adger 2006). Adaptation refers to the ability to respond to stresses (Nelson 2012). Importantly, "vulnerability is an approach often predicated on a moral and ethical responsibility to groups and populations that are more susceptible" (Nelson 2012: 376). A society's development path, physical exposure, resource distribution, social networks, government institutions and technology development influence its own vulnerability (IPCC 2007: 719–720). Without doubt, cities contain individuals and groups who are more vulnerable than others and lack the capacity to adapt to climate change (Schneider et al. 2007: 719).

Eventually, contemporary cities must develop a greater awareness of the need for policies that might eventually enhance resilience and reduce vulnerability to expected climate change impacts (Adger 2001; Vellinga et al. 2009). We need to map and draw the scenarios of uncertainties that may affect our cities as well as their communities and neighborhoods.

The multidisciplinary literature reveals that there are many frameworks and models for developing vulnerability assessment techniques (Adger 2006; Cutter et al. 2008; Fussel 2007; Green and Penning-Rowsell 2007; Manuel-Navarrette et al. 2007; McLaughlin and Dietz 2008). Cutter et al. (2008: 599) suggest that

© Springer Science+Business Media Dordrecht 2015   179
Y. Jabareen, *The Risk City*, Lecture Notes in Energy 29,
DOI 10.1007/978-94-017-9768-9_9

despite their differences, these approaches share some common features: (a) approaching vulnerability from a social-ecological perspective; (b) the significance of place-based studies; (c) the conceptualization of vulnerability as an equity or justice issue, and (d) the use of vulnerability assessment for identifying hazard zone aiming at pre-impact and hazard mitigation planning.

Yet, the existing multidisciplinary studies on urban vulnerability mostly focus on one or a limited type of risk or on a single specific hazard to assess their impacts (see Wilhelmi and Hayden 2010; Adger 2006). This chapter aims to produce a new framework and procedure for understanding and analyzing urban vulnerabilities, which I will name UVM—*Urban Vulnerability and Adaptation Matrix*. UVAM takes into consideration the fact our cities face multiple risks and various types of risk may occur solely or in various other combinations of risk.

9.2 Definition and Procedure

The *Urban Vulnerability Matrix* is a framework, or a model, for analyzing the social-spatial distribution of risk and vulnerabilities in a city and its neighborhoods or quarters on one hand, and the adaptation measures on the other hand. It provides us with significant information regarding risk and uncertainties at the level of city, communities, and social—groups. Eventually, the *Urban Vulnerability Matrix* helps us understand not only vulnerabilities at the risk city, but also to develop planning scenarios for the city as well as for every single neighborhood or section as needed. The concept of the Vulnerability Analysis Matrix is composed of four main components that determine its scope, environmental, social, and spatial nature.

The procedure of building the model of the *Urban Vulnerability Matrix* for a given city is based on nine steps as will see in the following pages. Moreover, I will use an example of a city in order to illuminate the procedure and use it for further theoretical insights.

The aim of this section is to illustrate an example for building UVM. The city that will be used as an example is Haifa, in northern Israel. It is the city in which I live and work, and I want to use its limited data in order to exemplify the constructing of UVM. Haifa is built on the slopes of Mount Carmel. The history of settlement in the area spans more than 3000 years. It is the home also to the Baha'i World Centre, a UNESCO World Heritage Site. Haifa is the largest city in northern Israel, and the central city in Metropolitan Haifa. It is located on the Mediterranean Sea. The city is rich with risk: it is a shore city, which will be affected by climate change and rising sea levels; it has a large industrial area with heavy and risky industries: petroleum refining and chemical processing; parts of its neighborhoods are adjacent to the Carmel Park which has experienced destructive forest fires; it is also located along an earthquake fault.

There are limited data regarding the probabilities of the occurrence of various type of risk. Therefore, I will use the existing data and keep the rest as illustration without concrete numbers. Studies on the vulnerability of Haifa and its metropolitan area are rare. My aim is to illustrate and exemplify the constructing of the UVM and not to address ruinously the case of Haifa. Assessment of the urban vulnerability such as flooding, rising of sea levels, earthquake and fire will be calculated as an inverse function of the distances that were calculated for each hazard.

In this case, urban vulnerability values were calculated considering a full probability of the natural hazards that were mentioned in the study case area. At the final stages of the construction of the UVM we are supposed to assign probabilities based on the maximum likelihood from historical data and the probability of occurrence of each hazard or scientific data or judgments of experts. Unfortunately, most of these data do not exist; therefore, I will add them figuratively.

Step 1. Building the sub-areas of the city

The first step of building the UVM is to compose the neighborhoods, districts, sub-areas, or statistical areas of the city. Each country and city has its own division of the city. Some name it statistical areas; others name it neighborhood or districts. This is the base layers of the UVM. We dismantle the city to its small unites of neighborhoods or other submission. This layer is the base and geographic reference of the rest of layers.

The layers are supposed to be built using GIS-Geographic Information System, which is capable of managing a large amount of spatially related information, integrate multiple layers of information, and addresses simple and complicated spatial quires. Many studies have used GIS for assessing urban vulnerability (Renard and Chapon 2010; Barczak and Grivault 2007).

Figure 9.1 presents Haifa's sub-areas, or neighborhoods of Haifa. These areas also called statistical areas. This layer represents the basic spatial setting of the analysis.

Step 2. Socio-economic city layers

The second step is to construct the socio-economic layer of the city. This layer indicates the socio-economic level of each neighborhood or sub-area. Socio-economic level or status is a statistical indicator that represents and summarizes a wide range of social and economic variables into a single variable. Some countries use a scale between 1 and 10, where 1 is the lowest socio-economic status, and 10 is the highest. When there is no one single indicator of socio-economic variable, we may use other variables like household income, or income per person.

The assumption behind the socio-economic and demographic layers is that the demographic and socio-economic variables affect the urban vulnerability. It assumes that there are individuals and groups within all societies who are more vulnerable than others and lack the capacity to adapt to climate change (Schneider et al. 2007: 719). Demographic, health, and socio-economic variables affect the ability of individuals and urban communities to face and cope with environmental risk and future uncertainties. These variables affect the mitigation of risk, response and recovery from natural disasters (Blaikie et al. 1994; Ojerio et al. 2010). Accordingly, many variables affect the vulnerability of individuals and communities. However, the main variables are income, education and language skills,

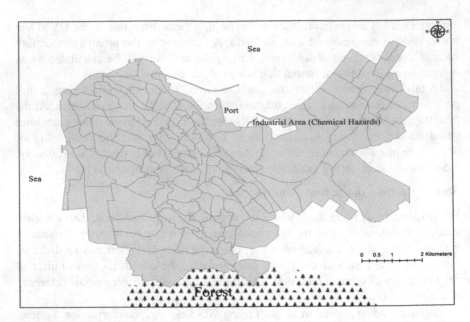

Fig. 9.1 Sub-areas or neighborhoods of Haifa

gender, age, physical and mental capacity, accessibility to resources and political power, and social capital (Cutter et al. 2003; Morrow 1999; Ojerio et al. 2010; United Nations Division for the Advancement of Women 2001). As a result, socio-economically weak communities are more vulnerable to suffer negative impacts, including property loss, physical harm, and psychological distress (Ojerio et al. 2010; Fothergill and Peek 2004).

Figure 9.2 demonstrates the socio-economic spatial settings of Haifa. It provides us with significant data regarding the poorest neighborhoods and the more affluent as well. Accordingly, the GIS analysis shows that concentration of the poor neighborhoods which are adjacent to each other.

Step 3. Demographic city layers

Building the demographic layers which include, for example: (1) age distribution, which will give us information about the more vulnerable groups such as children and the elderly; and (2) types of families: single parent family and others.

Figures 9.3 and 9.4 demonstrate the 'demographic layers', which include in our case the distribution of population by group age of 0–14 and 65 years and more in Haifa. It is assumed that these two age groups, 0–14 and 65 and more, are more vulnerable than others when a significant threat is taking place and happening. Some cities may have more detailed and significant demographic data that they should present and analyze.

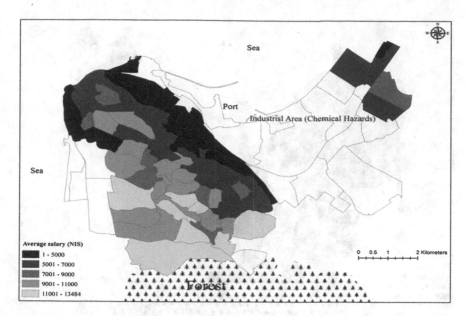

Fig. 9.2 Socio-economic layer of Haifa

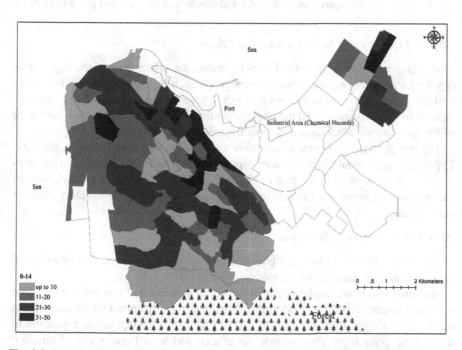

Fig. 9.3 Demographic layers: distribution of population by age 0–14 years in Haifa

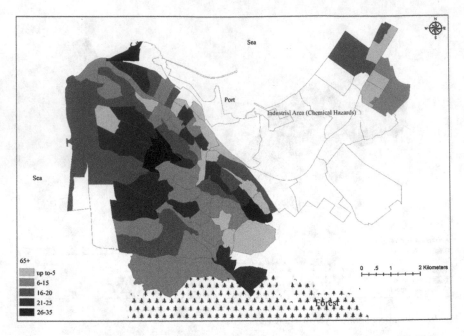

Fig. 9.4 Demographic layers: distribution of population by age 65 years and above in Haifa

Step 4. Typology city layers: urban conditions

This step aims to construct some layers regarding the urban form of the city and its typologies, such as housing densities, typographies, transportation networks, age of buildings, height of buildings, and so forth. These layers are informative and significant in order to understand the physical settings of the city, and to plan the future of the city and its adaptation measures by area or neighborhoods.

Figures 9.5 and 9.6 present the housing density and typography in the city. Typology maps should include more aspects of city form and physical typologies which they are missing in the case of Haifa. Yet, the housing density is a significant factor which is important to analyze mainly for evacuation and preventing risk in dense areas.

Step 5. Types of neighborhoods

Types of neighborhood by their physical and spatial conditions: formal-informal spaces; squatters; slums; and regular neighborhoods.

Informal spaces are unplanned, chaotic, and disorderly and it is assumed that the scale and human condition of informal places within a city have a significant impact on its vulnerability. According to UN-HABITAT (2008), much urban expansion in developing cities takes place outside the official and legal frameworks of building codes, land use regulations, and land transactions. Resilience requires the inclusion of the poor, vulnerable communities, and informal places in the city and in the

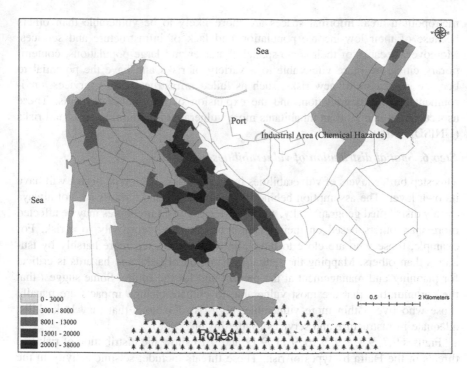

Fig. 9.5 Typology: housing density in Haifa

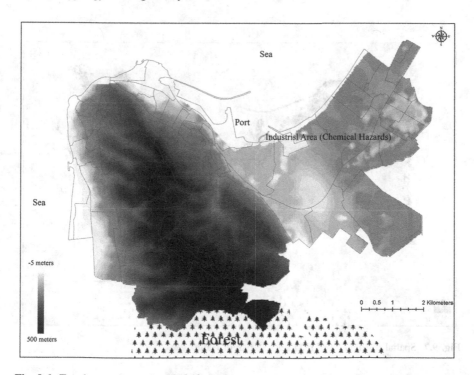

Fig. 9.6 Typology: typography of Haifa

metropolitan area. Informal spaces are more likely to be vulnerable than others because of their low-income population and lack of infrastructure and services. Moreover, because of their socio-spatial character and large populations, contemporary cities are more vulnerable to a variety of risks and have the potential to become generators of new risks, such as failed infrastructure and services, environmental urban degradation, and the expansion of informal settlements. These aspects make many urban inhabitants more vulnerable to natural hazards and risks (UNISDR 2010).

Step 6. Spatial distribution of vulnerabilities

This step builds layers of vulnerabilities by type of risk. Each type of risk will have its own layer. The assumption behind this is that risks and hazards are not always evenly distributed geographically, and some areas and communities may be affected more than others based on their location or intensity, sensitivity to a risk. For example, those who are close to the shore may be affected more harshly by tsunamis than others. Mapping the spatial distribution of risks and hazards is critical for planning and management at the present and for the future. Some suggest that the communities that are most vulnerable to climate change impacts are usually those who live within more vulnerable, high-risk locations that may lack skills, adequate infrastructure and services (Satterthwaite 2008).

Figures 9.7, 9.8, 9.9, 9.10, 9.11, 9.12 and 9.13 show the distribution of risk and threats in the Haifa by types of risk. These threats include: seismic activity in the city. It is important to mention here that buildings in Haifa, which were constructed

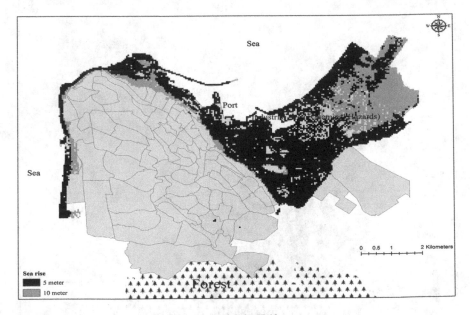

Fig. 9.7 Spatial distribution of risk: sea rise risk in Haifa

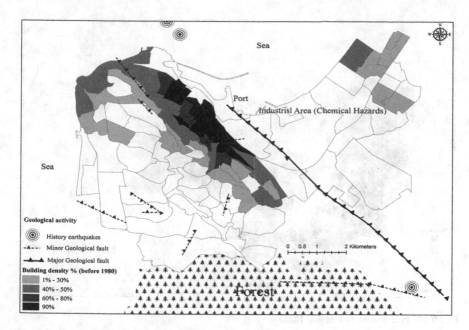

Fig. 9.8 Buildings at risk of seismic activity in Haifa

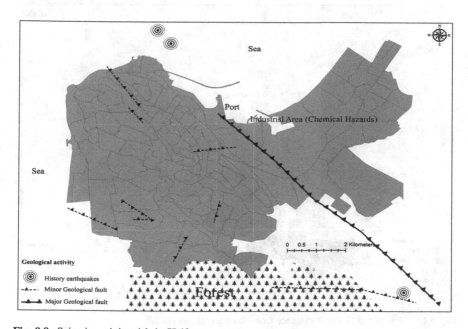

Fig. 9.9 Seismic activity risk in Haifa

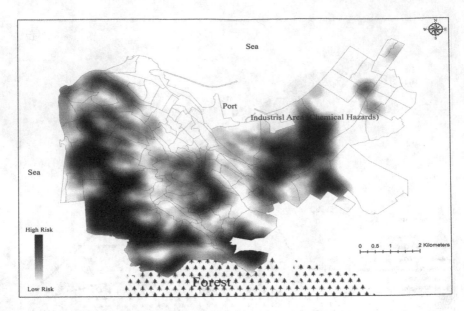

Fig. 9.10 Fire hazard risk inside Haifa

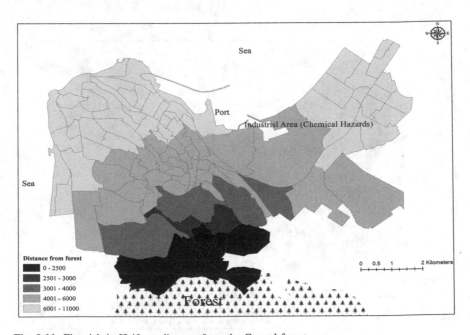

Fig. 9.11 Fire risk in Haifa as distance from the Carmel forest

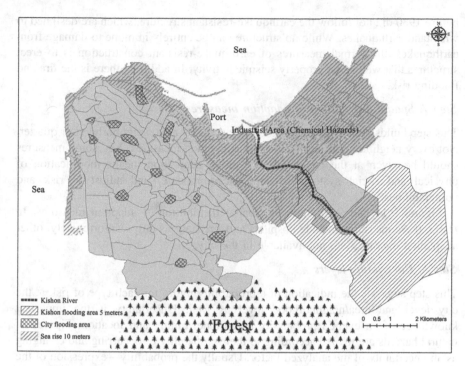

Fig. 9.12 Flooding risk in Haifa

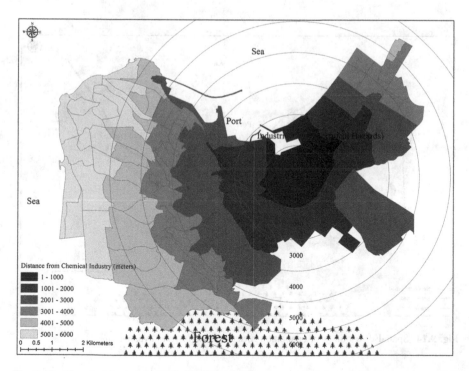

Fig. 9.13 Chemical hazard in Haifa by distance

before 1980 did not follow the earthquake-resistant structure, which are designed to withstand earthquakes. While no structure can be entirely immune to damage from earthquakes, the formal measures of earthquake-resistant construction is to erect structures that well resist properly seismic activity. In addition, there is the fire and flooding risk.

Step 7. Spatial distribution of adaption measures

This step builds the existing adaptation measures by area, neighborhood or quarter. Not every neighborhood needs the same adaptation measures. Yet, some measures should be given at the city level. Generally, adaptation means modification of physical and social systems to cope, accommodate, and adjust to risk and vulnerabilities.

Figures 9.14 and 9.15 show the spatial distribution of adaption measures. In Haifa, these measures include hospitals and fire stations only. Unfortunately, other adaptation measures are not available in the city.

Step 8. Uncertainty layers

This step suggests the indication of a probability figure for each type of risk at the city level and area/neighborhood level as well. Uncertainty is the imperfect knowledge of an event's probability magnitude, timing and location. Figures of natural hazards are treated according to their probability of occurring and changing as an uncertainty of the analyzed factor. Usually the probability—expression of the likelihood of an event occurring—of uncertainty is determined according to

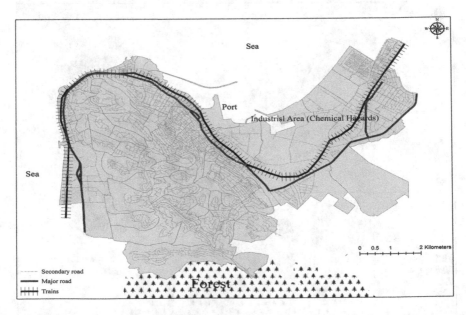

Fig. 9.14 Spatial distribution of adaptation measures: major road evacuation

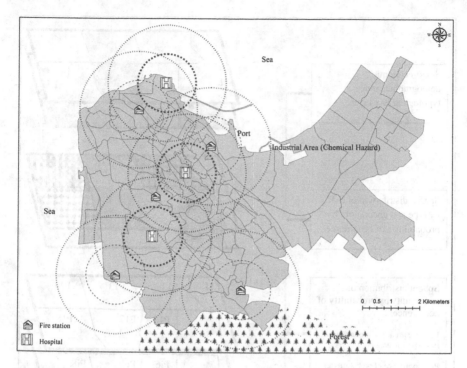

Fig. 9.15 Spatial distribution of adaption measures: hospitals and fire stations

historical data (e.g., maximum likelihood used to estimate to assess supply variability), (Means et al. 2010). The uncertainty probability will be considered for each Hazard Factor in the neural network structure.

Step 9. Vulnerability planning scenarios

Vulnerability, as well as risk, is differentially distributed in the city. Vulnerability planning scenarios are conceptualized and calculated for the city itself and for each neighborhood separately. Each area in the city has its own various risk and its own various sensitivity to risk. They also have their adaptation measures as well. They share many risk with the city, but in many cases, specific neighborhoods have their own specific vulnerability. By doing this, we provide residents of each area, neighborhood or quarter, the vulnerability conditions of their places (Figs. 9.16, 9.17, 9.18 and 9.19).

This chapter analyzes and identifies the types, demography, scope, and spatial distribution of environmental risk, and future uncertainties in a Haifa and its neighborhoods. Yet, the example shows that there are many missing data in Haifa. Similar to most cities in developing countries, the city does not have a climate change oriented plan; neither does it have a comprehensive risk assessment for the city and its neighborhoods. Moreover, there are limited data regarding the probabilities of the occurrence of various type of risk in Haifa, and there is a lack of

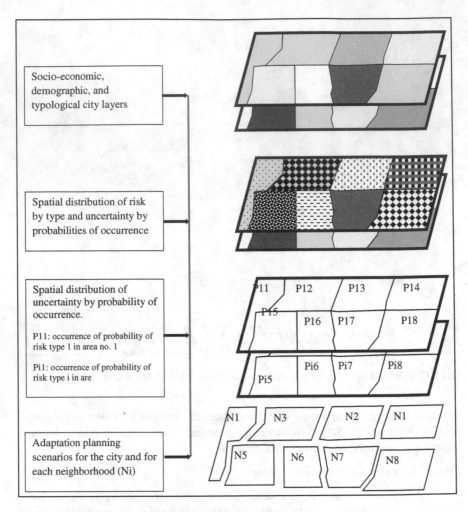

Fig. 9.16 The conceptual framework of urban vulnerability matrix

adaptation measures against flooding, sea rise, and increase of temperature in the city, fire hazards, chemical hazards and more. Ultimately, Haifa, like the vast majority of cities in the world, does not take the risk of its citizen and neighborhoods seriously. Yet, there are western cities that have begun to take the issues of risk city more seriously based on their deadly experience.

Our major conclusion regarding Haifa is that vulnerabilities in this city, like many other around the globe, are socially and spatially differentiated. Accordingly, low-income and poor neighborhoods are more vulnerable than others. Furthermore, their abilities to cope with risk are also lower than more affluent people and neighborhoods.

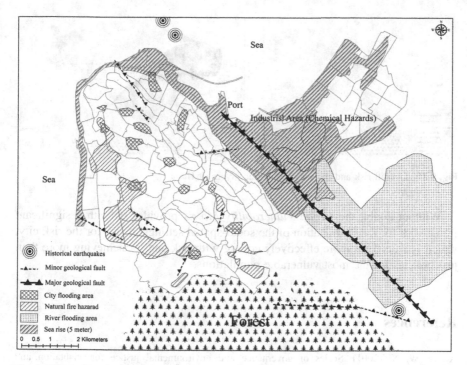

Fig. 9.17 Mapping overall risks at Haifa

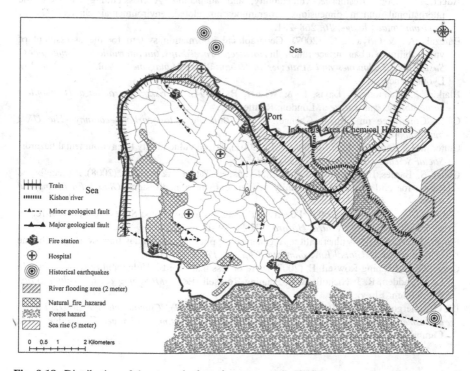

Fig. 9.18 Distribution of threats and adaptation measure in Haifa

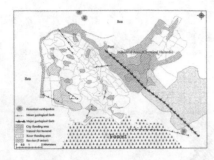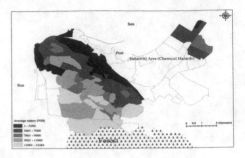

Fig. 9.19   Overall risk and the socio-economic layers of Haifa

Without doubt, the *Urban Vulnerability Matrix* provides us with a significant framework for the examination of the socio-spatial settings of risk for the risk city, which is crucial for more effectively coping with risk and for achieving more than just policies for the most vulnerable populations.

References

Adger, W. N. (2001). Scales of governance and environmental justice for adaptation and mitigation of climate change. *Journal of International Development, 13*(7), 921–931.

Adger, N. (2006). Resilience, vulnerability, and adaptation: A cross-cutting theme of the international human dimensions programme on global environmental change. *Global Environmental Change, 16*, 268–281.

Barczak, A., & Grivault, C. (2007). Geographical information system for the assessment of vulnerability to urban surface runoff. In *Novatech Proceedings, 6th International Conference—Sustainable Techniques and Strategies in Urban Water Management* (Vol. 1, pp. 31–146). Lyon, France.

Blaikie, P., Cannon, T., Davis, I., & Wisner, B. (1994). *At risk: Natural hazards, people's vulnerability, and disasters*. London: Routledge.

CCC—Committee on Climate Change. (2010). *Building a low-carbon economy—The UK's innovation challenge*. www.theccc.org.uk

Cutter, S. L., Boruff, B. J., & Shirley, W. L. (2003). Social vulnerability to environmental hazards. *Social Science Quarterly, 84*(2), 242–261.

Cutter, S., Barnes, L., Berry, M., Burton, C., Evans, E., Tate, E., & Webb, J. (2008). A place-based model for understanding community resilience to natural disasters. *Global Environmental Change, 18*, 598–606.

Fothergill, A., & Peek, L. (2004). Poverty and disasters in the United States: A review of recent sociological findings. *Natural Hazards, 32*(1), 89–110.

Fussel, H. M. (2007). Vulnerability: A generally applicable conceptual framework for climate change research. *Global Environmental Change, 17*(2), 155–167.

Green, C., & Penning-Rowsell, E. (2007) More or less than words? Vulnerability as discourse. In L. McFadden, R. J. Nicholls & E. Penning-Rowsell (Eds.), *Managing coastal vulnerability*. Amsterdam: Elsevier.

IPCC—Intergovernmental Panel on Climate Change. (2007). *Climate change 2007: Fourth assessment report of the intergovernmental panel on climate change*. Cambridge, MA: Cambridge University Press.

Manuel-Navarrette, D., Gomez, J. J., & Gallopin, G., (2007). Syndromes of sustainability of development for assessing the vulnerability of coupled human–environmental systems. The case of hydro meteorological disasters in Central America and the Caribbean. *Global Environmental Change, 17*(2), 207–217.

McLaughlin, P., & Dietz, T. (2008). Structure, agency and environment: Toward an integrated perspective on vulnerability. *Global Environmental Change, 18*(1), 99–111.

Means, E. III, Laugier, M., Daw, J., & Pirnie, M., Inc. (2010). *Decision support planning methods: Incorporating climate change uncertainties into water planning.* Prepared for: Water Utility Climate Alliance. Danver, CO: WUCA.

Morrow, B. H. (1999). Identifying and mapping community vulnerability. *Disasters, 23*(1), 1–18.

Nelson, D. R. (2012). Vulnerabilities and the resilience of contemporary societies to environmental change. In J. A. Matthews, P. J. Bartlein, K. R. Briffa, A. G. Dawson, A. De Vernal, T. Denham, S. C. Fritz & F. Oldfield (Eds.), *The sage environmental change* (pp. 374–386). London: Sage Publication.

Ojerio, R., Moseley, C., Lynn, K., & Bania, N. (2010). Limited involvement of socially vulnerable populations in federal programs to mitigate wildfire risk in arizona. *Natural Hazards Review, 12*(1), 28–36.

Renard, F., & Chapon, P. M. (2010). Using multicriteria method of decision support in a GIS as an instrument of urban vulnerability management related to flooding: A case study in the greater Lyon (France). *NOVATECH*, session 3.2. Available at http://documents.irevues.inist.fr/bitstream/handle/2042/35769/13208-111REN.pdf?sequence=1

Satterthwaite, D. (2008). *Climate change and urbanization: Effects and implications for urban governance.* Presented at UN Expert Group Meet. Popul. Distrib., Urban., Intern. Migr. Dev. UN/POP/EGMURB/2008/16/.

Schneider, S. H., Semenov, S., Patwardhan, A., Burton, I., Magadza, C. H. D., Oppenheimer, M., Pittock, A. B., Rahman, A., Smith, J. B., Suarez, A., & Yamin, F. (2007). Assessing key vulnerabilities and the risk from climate change. Climate Change 2007: Impacts, Adaptation and Vulnerability. In M. L. Parry, O. F. Canziani, J. P. Palutikof, P. J. van der Linden & C. E. Hanson (Eds.), *Contribution of working group II to the fourth assessment report of the intergovernmental panel on climate change* (pp. 779–810). Cambridge, UK: Cambridge University Press.

UN-HABITAT. (2008). *State of the world's cities 2008/2009—Harmonious cities.* London: Earthscan.

UNISDR-International Strategy for Disaster Reduction. (2010). *Making cities resilient: My city is getting ready.* 2010–2011 World Disaster Reduction Campaign.

United Nations Division for the Advancement of Women. (2001). Environmental management and the mitigation of natural disasters: A gender perspective. http://www.un.org/womenwatch/daw/csw/env_manage/documents/EGM-Turkey-final-report.pdf. July 7, 2009.

Vellinga, P., Marinova, N. A., & van Loon-Steensma, J. M. (2009). Adaptation to climate change: A framework for analysis with examples from the Netherlands. *Built Environment, 35*(4), 452–470.

Wilhelmi, O. V., Hayden, M. H., (2010). Connecting people and place: A new framework for reducing urban vulnerability to extreme heat. *Environmental Research Letters, 5,* 1–7.

Chapter 10
Conclusions

Though cities have always been "risk cities," this book argues that the cities of the contemporary postmodern world currently face a myriad of risks of unprecedented magnitude. For this reason–like contemporary societies, which are characterized by "the inherent pluralization of risks" they face (Beck 1997, p. 32)—our contemporary cities must also be understood in light of this constitutive concept. The already existing risks and constantly emerging new risks faced by the contemporary risk city have profound influence over urban social form and politics, from individuals and households to formal institutions and civil society. Risk, therefore, must be understood as a major force driving change and social transformation in urban societies. That being the case, it must also feature as a decisive concept in the theory and practice of urban planning. Moreover, risk, as a negative resource, is socially and spatially differentiated in the risk city, and therefore, it becomes a major concept of social and spatial inequality in cities.

10.1 The Evolution of the Risk City and Its Related Planning Practices

This book has emphasized the manner in which emerging risks continue to challenge the concepts, procedures, and scope of existing planning theory and practice. One fundamental premise of the risk city is that change in risk perception can lead to change in trust perception, both of which inform and induce the need for new practices to meet emerging challenges. Moreover, because the risk city is an evolutionary process that turns on the intertwining dialectical relationships among risk, trust, and practice, a change in risk perception can be expected to bring about fundamental change in the risk, trust, and practice constructs of the risk city and, therefore, in the overall setting of the risk city as a whole, as reflected in Fig. 10.1.

The threats faced by cities change over time, requiring them to employ different practices in order to provide their residents with refuge, protection, and a sense of trust and security. As these changing threats and uncertainties also help shape planning practices, particular chronological periods characterized by specific

© Springer Science+Business Media Dordrecht 2015
Y. Jabareen, *The Risk City*, Lecture Notes in Energy 29,
DOI 10.1007/978-94-017-9768-9_10

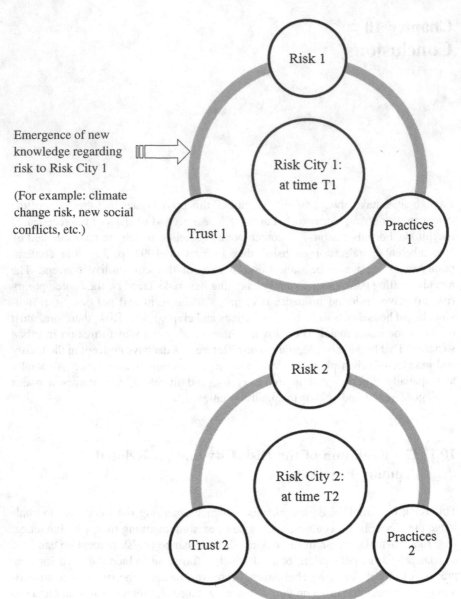

Emergence of new
knowledge regarding
risk to Risk City 1

(For example: climate
change risk, new social
conflicts, etc.)

Fig. 10.1 The evolutionary process of the risk city

conditions of urban risk and vulnerability witness the emergence of unique city
planning movements. From this perspective, the risk city as a conceptual framework allows us to better understand the transformation of planning theories,
movements, practices, and systems in different historical contexts.

This is particularly true of the many modern approaches to planning that have emerged since the late 19th century. Modern planning emerged at the turn of the 20th century in response to the profound crisis in urban organization and function and the urban social problems engendered by these phenomena, from the hardships and impoverishment of the working class to the congestion, physical degradation, and functional chaos of the modern city (Beauregard 1989; Gans 1968; Hall 2000; Harvey 1989). This crisis posed a fundamental risk to the economic and social functioning of cities. To meet this challenge, planners and other practitioners since the early 20th century have proposed different models and theoretical and practical approaches aimed at remedying modern urban malaise, reconstructing and reshaping modern social space, and addressing the problem of sprawl in the United States and other developed countries. One example is the "Garden City Movement," which arose in England in response to the crowding and pollution of cities resulting from the Industrial Revolution. According to Ebenezer Howard's *To-Morrow: A Peaceful Path to Real Reform* (1898), the idea of establishing new towns was advanced as an alternative to the threats then facing the cities of England. Another example from the final decade of the 1800s was the "City Beautiful Movement," which took form in the United States in reaction to urban malaise, poverty, crime, congestion, blight, and ugly repetition (Borbely 2007).

These are just a few examples of the innovative planning initiatives that have been advanced since the second half of the 19th century to address the social and spatial threats facing cities. Others include Haussmann's project in Paris in the 1860s; the "Great White City" model presented at the Chicago World's Fair of 1893; Le Corbusier's "City of Tomorrow" and "Plan Vision" for Paris of 1924; Perry's Neighborhood Unit (Perry 1939), first advanced in the late 1920s; and the large-scale public housing and urban renewal initiatives implemented in the 1950s and 1960s in the wake of WWII (Harvey 1989). Among other things, these initiatives were based on the premise that urban social ills could be mitigated through the promotion of beauty, which, it was believed, would inspire residents to higher civility and morality (Borbely 2007). Indeed, by the mid-19th century, it was evident that economic logic had repressed alternative social rationalities and produced "anti-human landscapes" (Gleeson 2000). In this context, planning emerged as a modernizing force aimed at enlightening, both metaphorically and literally, the murky inhuman spaces produced by the market (Gleeson 2000). Early planning literature–such as the writings of Fitzgerald (1906), Sulman (1921), and Barnett et al. (1944)—is reflective of this desire to bring light to cities that had been darkened by the risks produced by the market (Gleeson 2000; Gleeson and Low 2000). In his *Cities of Tomorrow*, Hall (2000: 48) argues that "most of the philosophical founders of planning movement continued to be obsessed with the evils of the Victorian slum city—which indeed remained real enough, at least down to World War Two, even to the 1960s".

Many of the current approaches to planning and architecture crystallized in the wake of the rise of environmental consciousness in recent decades, in response to contemporary critical social and environmental concerns. These approaches include new-urbanism, transit-oriented development, neotraditionalism, the "urban village," and the "transit village"; various models of sustainability, such as the "sustainable city," the "sustainable community," the "health city," and the "green city"; and the rise of plans geared specifically toward climate change.

10.2 The Theory and Practice of Planning for Countering Climate Change

As we have seen, the continually emerging conditions of risk, stemming primarily from the phenomenon of climate change and its resulting uncertainties, challenge the concepts, procedures, and scope of existing approaches to city and community planning and ultimately result in new practices and new settings of risk and trust perceptions among the residents and decision makers of our contemporary cities. This book advances the conceptual framework of Planning for Countering Climate Change (or PCCC)—a *praxis* that synthesizes the knowledge and skills necessary to effectively manage and cope with climate change in the urban context. PCCC differs from traditional and conventional planning approaches in its data analysis, visioning, procedures, and practices. The planning practices of PCCC are informed by concepts related to the constantly emerging risks of climate change; by demographic, economic, and spatial analysis; and by analysis of the risk and uncertainties facing cities today. In this way, they recognize and embrace the fact that knowledge regarding the impacts of climate change and efforts to contend with them have become a major resource for spatial planning. PCCC incorporates planning concepts that are not central to conventional planning approaches, such as adaptation and mitigation policies, energy, ecological and green economics, urban risk mapping and assessment (the Urban Vulnerability Matrix), and public and expert involvement. The adaptation measures employed by PCCC are based on the premise that planners must also think in terms of "urban defensibility," or protection of the city. PCCC's concern with energy as a major guiding concept in the planning of cities and communities is also inextricably linked not only to sustainability efforts but to issues of climate change. Also on the practical level, PCCC applies scenario-planning to help explore policy options regarding where to build, what to build, and how to strengthen communities in the areas of greatest risk.

PCCC also provides the basis for a new conceptual framework encompassing easy to grasp methods for evaluating urban plans: the Countering Climate Change Evaluation Method (CCCEM). CCCEM acknowledges the essentially qualitative nature of urban phenomena and contributes to the scholarship by applying an innovative, multidisciplinary qualitative methodology to complexity theories in countering climate change in the urban context. Because it is easy to grasp and makes intuitive sense, CCCEM has the potential to promote greater awareness among scholars, professionals, decision makers, and the general public regarding the current and future direction of cities as far as climate change issues are concerned.

10.3 Contemporary Planning of the Risk City in New York City

This book also contains a CCCEM-based analysis of the planning efforts to counter the impacts of climate change in New York City, as reflected in the recent ambitious plan for New York City: *PlaNYC 2030*. Our analysis of the plan itself (as opposed to other activities related to climate change that may be underway), which is actually a strategic plan that aims to counter climate change in the city, seems to indicate that New York City takes the risk of climate change seriously. Climate change was a major concern in the formulation of the plan's problem, justification, visioning, and objective. *PlaNYC* applies an integrated planning approach that makes use of the advantages of new urbanism, T.O.D., sustainable development, mitigation efforts, and the monitoring of institutional policies.

PlaNYC is a physically oriented plan that focuses primarily on the reconstruction of infrastructure, the promotion of greater compactness and density, the enhancement of mixed land use, sustainable transportation, greening, and the renewal and utilization of empty parcels and brownfields. *PlaNYC* also advances an ambitious vision of a 30 % reduction in emissions and a "greener, greater New York," and links this vision to the international discourse on climate change and the international climate change agenda.

Still, *PlaNYC* fails to adequately address the social issues that are so crucial to New York City as one of the most diverse cities in the world. It also fails to address the climate-change related issues facing vulnerable communities. Indeed, our analysis found that New York City is socially differentiated in the capacity of its communities to contend with climate change uncertainties, physical and economic impacts, and environmental hazards. PlaNYC also failed to effectively integrate civil society, communities, and grassroots organizations into the planning process. The lack of a systematic procedure for public participation throughout the city's neighborhoods and among different social groupings and stakeholders reflects a critical shortcoming of the planning process, particularly in the current age of climate change uncertainty. Finally, *PlaNYC* failed to make a sufficiently radical shift toward planning for climate change in its failure to propose adequate adaptation measures—an assessment that was confirmed by the catastrophic outcome of Hurricane Sandy.

10.4 The Appalling Test of Hurricane Sandy

Hurricane Sandy provides us with a significant opportunity to examine the resilience of New York City and to draw conclusions regarding how the city might better work in the future to confront the impacts and hazards of climate change. Overall, although New York City has a plan for countering climate change impact and has begun implementing it, the city has thus far proven unable to truly prepare itself for the serious risks that lie ahead. The unfortunate impact of Hurricane Sandy reflects the marked lack of resilience of the current institutional and spatial settings of our cities. As a result, it also reflects the fact that our cities have become risky places for their residents during hazardous events. The critical task of the city at present is to prepare for the uncertainties of the future. To this end, city officials must learn and implement the lessons of Hurricane Sandy in order to build the future of the city and promote its resilience for all. As wisely asserted in the forward to the recommendations of the NYS 2100 Commission, "we cannot just restore what was there before—we have to build back better and smarter."

10.5 The Risk City and Its Resilience Framework

One response to the contemporary risk city has been the emergence of the concept of resilient cities and an extensive body of literature on the subject. Borrowed from ecology and the sciences, this concept has been incorporated into urban studies, planning, and other related fields that grapple with issues of climate change at the city level. In a previous chapter, I introduce the Risk City Resilience Trajectory, which is based on the premise that, since "resilience requires frequent testing and evaluation" (NYS 2100 2013: 7), our cities must learn from the past and the present in order to plan for the uncertainties of the future. In this context, learning should be based chiefly on our experience and emerging knowledge on vulnerability and adaptation measures. According to the Resilient City Framework, a resilient city is defined by the overall ability of its physical, economic, social, and governance systems, and other entities that are exposed to hazards, to learn from, effectively plan and prepare for, resist, absorb, accommodate, and recover from the effects of a hazard in a timely and efficient manner, including through the preservation and restoration of its essential basic structures and functions. The Risk City Resilience Trajectory requires acknowledging current and future vulnerabilities and risks in order to plan a different future and is based on the conviction that we need to plan, build, and reconstruct our cities in a different and smarter manner. The Risk City Framework is a dynamic and flexible framework that acknowledges the complexity of city resilience and its non-deterministic and uncertain nature.

10.6 Planning Practices Around the World

Every society and every city understands differently the risks it faces. In this book, I have attempted to show how changes in urban risk perception has resulted in the production of different planning practices and initiatives in cities around the world, primarily in Western Europe, North America, and Australia. As plans have the ability to incorporate mitigation, adaptation, land-use, energy, social, and economic policies, all within one integrated framework, city plans have become exceedingly significant instruments for coping with risk in general and the risk stemming from climate change in particular. The question, however, is whether such planning adequately addresses the present and future risks and uncertainties facing city residents. Based on the qualitative assessment of our international sample of developed and underdeveloped cities, the current situation appears to be quite bleak. Our analysis suggests that our cities are neither properly nor effectively fulfilling the critical role they should be playing in coping with climate change risk and the uncertainties facing their own residents. Some cities have used their plans to articulate their view that climate change and environmental hazards are major risks with which they must contend, whereas other cities, which I assume are representative of the vast majority of cities around the world (in Russia, China, and other developing countries), appear to perceive risk differently—that is, as related not to climate change but to the importance of effectively exploiting future growth opportunities.

The quest for city growth and uncontrollable profit (reflecting the dominant trend in most cities around the world), combined with neoliberal urban agendas in which the city becomes a negotiable object on the competitive market, leads decision makers at the national and local government levels to neglect fundamental issues of threat and risk. As a result, our contemporary cities are not doing all they can to fortify themselves against uncertainties, climate change, and natural and environmental hazards, and for this reason they may prove to be deathtraps for millions of residents when disasters occur.

The major concerns of developing cities around the world are not related to climate change but rather focus on fundamental human needs of their own population, such as food, clean water, urban hygiene, shelter, and employment. Rarely, therefore, do such cities employ building adaptation and mitigation policies aimed at coping with climate change issues.

10.7 Inequality in the Risk City

As negative resource risk is unequally distributed among social groups and neighborhoods, risk itself is socially differentiated. For this reason, coping with risk using means such as adaptation policies (for example) is also unequally distributed. Although, some types of risk and threats are equally distributed in the city, poor and disadvantaged people and neighborhoods are less likely to be able to cope adequately with various risks. A socio-spatial mapping of vulnerabilities in the risk city using the Urban Vulnerability Matrix reveals that in many cities around the world, vulnerabilities are socially and spatially distributed in an unequal manner. The risk city often becomes a victim of the neoliberal economy and the devastating unlimited growth and profits it engenders. However, the risk city also simultaneously possesses the power to challenge the neoliberal agenda of cities based on the premise that the lives of millions of urban inhabitants literally hang in the balance.

References

Beauregard, R. A. (1989). Between modernity and postmodernity: The ambiguous position of US planning. *Environment and Planning D: Society and Space, 7(4),* 381–395.

Beck, U. (1997). *The reinvention of politics: Rethinking modernity in the global social order* (M. Ritter, Trans.). Cambridge: Polity.

Borbely, M. (2007). *Residence parks: An American vision reborn.* Summer: American Bungalow Magazine.

Barnett, O., Burt, W. O. & Heath, F. (1944) *We must go on: A study in planned reconstruction and housing.* Melbourne: The Book Depot.

Fitzgerald, J. D. (1906). *Greater Sydney and greater Newcastle.* Sydney: New South Wales Bookstall.

Gans, H. J. (1968). *People and plans: Essays on urban problems and solutions.* New York: Basic Books.

Gleeson, B. (2000). Reflexive Modernization: The Re-enlightenment of Planning? *International Planning Studies, 5(1):* 117–135.

Gleeson, B. & Low, N. (2000). *Australian Urban Planning: New Challenges: New Agendas.* Allen and Unwin: Crown Nest.

Hall, P. (2000). *Cities of tomorrow.* Malden, MA.: Blackwell.

Harvey, D. (1989). *The condition of postmodernity.* Oxford: Blackwell.

Perry, C. (1939). *Housing for the machine age.* New York: Russell Sage Foundation.

Rodin, J., & Rohaytn, F. G. (2013). *NYS 2100 commission: Recommendations to improve the strength and resilience of the empire state's infrastructure.*

Sulman, J. (1921). *Town planning in Australia.* Sydney: Government Printer.

Printed in the United States
By Bookmasters